EAI/Springer Innovations in Communication and Computing

Series Editor
Imrich Chlamtac, European Alliance for Innovation, Ghent, Belgium

The impact of information technologies is creating a new world yet not fully understood. The extent and speed of economic, life style and social changes already perceived in everyday life is hard to estimate without understanding the technological driving forces behind it. This series presents contributed volumes featuring the latest research and development in the various information engineering technologies that play a key role in this process. The range of topics, focusing primarily on communications and computing engineering include, but are not limited to, wireless networks; mobile communication; design and learning; gaming; interaction; e-health and pervasive healthcare; energy management; smart grids; internet of things; cognitive radio networks; computation; cloud computing; ubiquitous connectivity, and in mode general smart living, smart cities, Internet of Things and more. The series publishes a combination of expanded papers selected from hosted and sponsored European Alliance for Innovation (EAI) conferences that present cutting edge, global research as well as provide new perspectives on traditional related engineering fields. This content, complemented with open calls for contribution of book titles and individual chapters, together maintain Springer's and EAI's high standards of academic excellence. The audience for the books consists of researchers, industry professionals, advanced level students as well as practitioners in related fields of activity include information and communication specialists, security experts, economists, urban planners, doctors, and in general representatives in all those walks of life affected ad contributing to the information revolution.

Indexing: This series is indexed in Scopus, Ei Compendex, and zbMATH.

About EAI - EAI is a grassroots member organization initiated through cooperation between businesses, public, private and government organizations to address the global challenges of Europe's future competitiveness and link the European Research community with its counterparts around the globe. EAI reaches out to hundreds of thousands of individual subscribers on all continents and collaborates with an institutional member base including Fortune 500 companies, government organizations, and educational institutions, provide a free research and innovation platform. Through its open free membership model EAI promotes a new research and innovation culture based on collaboration, connectivity and recognition of excellence by community.

Manolo Dulva Hina • Amar Ramdane-Cherif •
Rafik Zitouni • Assia Soukane
Editors

Advances in Computational Intelligence and Communication

Selected Papers from the 2nd EAI International Conference on Computational Intelligence and Communications (CICom 2021)

Springer

EAI
RESEARCH MEETS INNOVATION

Editors
Manolo Dulva Hina
SIC Laboratory, Omnes Education Research Centre
ECE Paris School of Engineering
Paris, France

Amar Ramdane-Cherif
LISV laboratory
University of Versailles - Paris-Saclay
Vélizy, France

Rafik Zitouni
5GIC & 6GIC, Institute for Communication Systems (ICS)
University of Surrey
Guildford, UK

Assia Soukane
ECE Research Centre
ECE Paris School of Engineering
Paris, France

ISSN 2522-8595 ISSN 2522-8609 (electronic)
EAI/Springer Innovations in Communication and Computing
ISBN 978-3-031-19522-8 ISBN 978-3-031-19523-5 (eBook)
https://doi.org/10.1007/978-3-031-19523-5

This Springer imprint is published by the registered company Springer Nature Switzerland AG
The registered company address is: Gewerbestrasse 11, 6330 Cham, Switzerland

Preface

This book contains some selected papers from *CICom 2021, the second EAI International Conference on Computational Intelligence and Communications*. The book is intended as academic and research material for students, educators and researchers who wish to learn the latest developments in the domain of computational intelligence and communications.

Artificial Intelligence is a domain that deals with the design, development and implementation of various systems capable of imitating human intelligence. *Computational intelligence* is a subset of artificial intelligence. It is a field of computer science by which a computer learns some specific tasks using learned data or experimental observations. It addresses complex real-world problems in which using mathematical or traditional modelling is not the preferred solution given that (i) the involved computational process might be too complex for mathematical reasoning; (ii) it might contain some uncertainties during solution processing; or (iii) by nature, the computational process is a randomly determined one (heuristic). Nature-inspired phenomena are imitated, and different solutions inspired by such phenomena are adapted. Nowadays, computational intelligence has evolved to include neural networks, fuzzy logic systems and evolutionary computation, as well as social reasoning, artificial life, ambient intelligence, deep learning and the likes. This book contains eight chapters, each of which presents typical topics and problems in which various computation intelligence solutions are adapted.

Chapter 1 is about the perception enhancement of a vehicle in a bad weather environment. Bad weather due to rain, fog or reduced visibility hinders the proper functioning of a sensor. The chapter proposes the use of drones (a.k.a., UAV or unmanned aerial vehicle) to supplement or enhance the vehicle's perception of its environment. The environment perception and vehicle/UAV interactions are managed by a knowledge base in the form of an ontology, and logical rules are used in order to detect and infer the environmental context and the UAV management. A use case is presented, tested and validated using a simulator developed using the Unity 3D programming platform.

Chapter 2 presents a work on adaptive neural backstepping (Backstepping RBF) to control three joints (hip, knee and ankle) of a right lower limb exoskeleton. To validate its proposed controller, cosimulation experiments were performed using SolidWorks, SimMechanics and MATLAB/Robotics Toolbox. The results showed that the authors' proposed approach produces the best performance compared to the backstepping controller, and the backstepping 1st and 2nd order sliding mode controller techniques. The proposed method ensures tracking of the desired trajectory with high accuracy while it generates a torque without chattering.

Chapter 3 deals with an evaluation framework for blockchain-based public key infrastructures. The X.509 Public Key Infrastructure (PKI) is an important trusted infrastructure for establishing a secure connection among participants on the Internet. Nevertheless, the infrastructure is far from perfect since various attacks in the past decade successfully exploited its weaknesses. A common idea for the creation of a more resilient PKI is the use of distributed ledger technology such as blockchain. In this chapter, the authors propose a set of requirements which can be used to evaluate and compare the merits of a certain design. This framework is put to the test by a review of twelve published blockchain-based PKI replacements or extensions. The authors recommended integrating the most promising ideas put forward by different proponents. In doing so, a holistic and scalable solution becomes possible.

Chapter 4 deals with evaluation of the performance of MIMO vehicular visible light communications. Vehicular communication allows connectivity and data sharing between traveling vehicles, becoming a tool to improve road safety and comfortable driving. In this chapter, the performance of Multiple-Input Multiple-Output (MIMO) vehicular communication system utilizing the visible light communication (VLC) technology is investigated. The chapter considered a VLC-based-vehicle-to-vehicle (V2V) system, where two vehicles follow each other on a single-lane road. The received optical power and the bit error rate (BER) are investigated for each MIMO link between transmitters and receivers. The impact of displacements, weather conditions and receiver apertures is also discussed.

Chapter 5 proposes an efficient big data architecture based on microservices. Software architectures for big data generally correspond to a stack of software chosen initially for their functionality and compatibility. Big Data applications evolve at different rates. The execution of these applications requires configurations that depend on the resources used. These configurations are not fixed; some evolve over time. The authors looked back at their Big Data architecture to make it more manageable and more reactive to the execution context by proposing a Kubernetes orchestration based on business microservices. As a result, the use of computing resources has improved, and data access became better managed. Their application log management strategy used the distribution of computational case by separating the activity monitoring from the big data computation code.

Chapter 6 deals with an intelligent target cell selection algorithm for low latency 5G networks. The fifth generation (5G) networks are characterized by ultra-dense deployment of base stations with a limited footprint. Consequently, user equipments handover frequently as they move within 5G networks. In addition, 5G requirements of ultra-low latencies imply that handovers should be executed swiftly to minimize service disruptions. To preserve security and privacy while maintaining optimal performance during handovers, numerous schemes have been developed. However, majority of these techniques are either limited to security and privacy or address only the performance aspect of the handover mechanism. For this reason, there is need for a novel handover authentication protocol that addresses security, privacy and performance simultaneously. This chapter developed a machine learning protocol that not only facilitated the optimal selection of target cell but also upheld both security and privacy during handovers. The authors' simulation results showed that their protocol had lower packets drop rate and pingpong rate, with a higher ratio of packets received compared to improved 5G authentication and key agreement (5G AKA') protocol. In addition, the proposed protocol was robust against packet replays, eavesdropping, forgery, masquerade, man-in-the-middle (MitM) and denial of service (DoS) attacks.

Chapter 7 presents "Advances in Coyote Optimization Algorithm: Variants and Applications". The Coyote Optimization Algorithm (COA) is a recent population-based algorithm inspired by the intelligent social behaviour of coyotes. COA has been widely applied to tackle different optimization issues in several areas and has proven its success compared to other meta-heuristics found in the literature. In this chapter, the authors describe an overview of COA and its variants including modified and hybridized versions. Moreover, COA applications in various fields such as image segmentation, wireless sensor networks, economic dispatch, electric power systems, distributed generation and other engineering problems are presented. Finally, we recommend some interesting future research areas directions for COA.

Chapter 8 presents "K Semantics for Security Policy Enforcement on Android Applications with Practical Cases". Android is one of the most widely used operating systems, yet android applications have helped malware creators to break in and infect smartphone devices. A long list of threats caused by downloaded applications targets the integrity of the system and the privacy of users. To protect android users from such applications, the authors propose a formal environment for security policy enforcement. The main idea is to establish a synergy between the aspect-oriented paradigm and formal methods such as the program rewriting technique. The approach is to rewrite the application program by adding control tests at certain carefully selected points to ensure that the security policy is respected. The new version of the program preserves all the good behaviours of the original version, complying with security policy and acting against the bad ones. Here, all the steps for security policy enforcement are performed automatically using the K-framework. This semantic engineering framework handles the whole process, from the formal specification of the policy and the program up to the enforcement technique. Two practical use cases are discussed in the chapter.

The chapters presented above are indeed latest research works in the domain of recent advances in computational intelligence and communications.

Paris, France — Manolo Dulva Hina
Versailles, France — Amar Ramdane-Cherif
Surrey, England, UK — Rafik Zitouni
Paris, France — Assia Soukane

Contents

Vehicular Perception Improvement in an Inclement Weather Context

Abderraouf Khezaz, Manolo Dulva Hina, Hongyu Guan, and Amar Ramdane-Cherif

1 Introduction

There are multiple traits characterizing an autonomous vehicle. One of many definitions, based on a legal context, describes an autonomous vehicle as a "vehicle that uses artificial intelligence, sensors, global positioning system coordinates, or any other technology to carry out the mechanical operations of driving without the active control and continuous monitoring of a human operator," [20] effectively implying that it uses both software and hardware elements to perceive and safely navigate its surroundings. It also suggests that sensors are the most significant additional component of an autonomous vehicle, as they serve as the vehicle's perception tool.

Sensors are electronic components that can only function under certain conditions. By its very nature, an autonomous vehicle must drive through a variety of surroundings, and differences in brightness and weather may have an impact on sensors performances.

A. Khezaz (✉)
ECE Paris, Paris, France

Université de Versailles-Saint-Quentin-en-Yvelines, Laboratoire d'Ingénierie des Systèmes de Versailles, Vélizy-Villacoublay, France
e-mail: abderraouf.khezaz@ece.fr

M. D. Hina
ECE Paris, Paris, France
e-mail: manolo-dulva.hina@ece.fr

H. Guan · A. Ramdane-Cherif
Université de Versailles-Saint-Quentin-en-Yvelines, Laboratoire d'Ingénierie des Systèmes de Versailles, Vélizy-Villacoublay, France
e-mail: hongyu.guan@lisv.uvsq.fr; rca@lisv.uvsq.fr

M. D. Hina et al. (eds.), *Advances in Computational Intelligence and Communication*, EAI/Springer Innovations in Communication and Computing,
https://doi.org/10.1007/978-3-031-19523-5_1

In general, autonomous vehicles are evolving in an environment that is becoming more intelligent and networked, and the concept of a *"smart city"* is slowly taking shape[26]. It relies on the ability of multiple entities to communicate in order to exchange information and maintain security for the inhabitants. It would be interesting to use this to improve the perception of the car in a situation where it is required. For example, vehicle sensors have been demonstrated to be ineffective in bad weather, but new technology, such as unmanned aerial vehicles (UAVs, or Drones), may be able to alleviate this problem.

This study proposes a knowledge-based model that recognizes a hazardous weather condition and employs a UAV and logical principles to ensure vehicular security. The work is then evaluated in a driving simulator.

Structure of the paper: Sect. 2 is dedicated to the Related Works. Section 3 describes the ontology developed for this work. Section 4 introduces the simulator and the use case used for testing. The paper is finally concluded by an analysis and a perspective of future works.

2 State of the Art

The weather is an uncontrollable external variable. The quality of the weather can significantly reduce a driver's visibility, and this phenomenon has been extensively researched in the past, with some studies dating back to the 1970s [6]. Harith et al. [13] made a review of 45 distinct works covering more than 500,000 accidents due to unfavorable weather, particularly rain and fog conditions. They come to the conclusion that driver alertness is the most important factor in keeping the road safe.

Das et al. [9] concur that rainy weather is one of the most dangerous driving conditions, generating up to 25% of crashes in some areas. They put together a dataset based on Florida crash records and were able to show a link between poor vision due to weather events and traffic accidents.

The same group employed data mining on larger data sets in order try to find aggravating factors that potentially raise the accident rate when combined with rain [10]. The pattern that emerges can provide significant insight for safety professionals, but a comparison with a clear-weather dataset was not available at the time of the study.

According to Andrey et al. [1], even if the accident rate rises by up to 70% when it rains, it returns to normal when the rain ends, despite the lingering effect of wet roads. This implies that the low visibility conditions, rather than the slick roadways, may be the primary cause of those incidents.

In light of the findings of the studies on the subject, it is apparent that bad weather and low visibility have a negative impact on road safety. When it comes to human drivers, this should be addressed by an increased awareness and an attempt to improve global visibility(fog lamps, windshield wipers...). However, in an Intelligent Transportation System (ITS), it is mainly about a general improvement of perception [32]. Van Brummelen et al. [29] conducted an extensive review of the current state of vehicular perception in their 2018 study. Perception, Localization and Mapping, Path Planning, Decision Making, and Vehicle Control are the five main components of autonomous vehicle navigation, with Perception being defined as using "sensors to continuously scan and monitor the environment, similar to human vision and other senses." To do this, a large number of different sensors should be considered [7, 29, 30]:

- Radars have been utilized for vehicle applications for decades, according to Woll [31]. This technique has proven to be excellent in mid-to-long-range measurement and precision, as well as performing well in adverse weather conditions [24]. It is still widely used in automobiles, but it has a limited field of view and performs poorly in near-distance measuring and static object identification. There is also the possibility of getting a noise signal from other sources or cars.
- LIDAR technology relies on measuring laser light reflection to infer the distance to a target. It has been studied since the 1980s[25], but it was not until the early 2000s that it was used in a vehicle [18]. It is a valuable tool for 3D mapping and localization, and it may be used on a large FOV [24], but it is very dependent on good weather and is ineffective outside of a specific range.
- Infrared cameras detect moving objects by measuring temperature radiations, and they perform well in both bad weather (rain, snow, fog) and low light conditions [15]. They cannot, however, be used to classify or determine the distance between objects in a picture.

Other types of sensors, such as ultrasounds, can be found aboard automobiles. However, no single "ideal" sensor exists that can provide a flawless perception in a bad weather.

Bengler et al. [4] remarked in their 2015 ADAS evaluation that vehicular perception has progressed from being concentrated on the vehicle to its surrounds, and that the next natural step will be to combine data from various sensors to improve dependability.

The usage of a sensor that is external to the vehicle would then be interesting to examine. The UAV (Unmanned Aerial Vehicle), commonly known as a Drone, is one such tool. Menouar et al. [19] offered the notion of using UAV as supporting components to ITS (Intelligent Transportation System) in their 2017 article, suggesting different uses. Indeed, in a world where transportation is primarily two-dimensional, their ability to travel in 3D space at fast speeds, as well as their size, which allows package transportation while remaining smaller than automobiles, gives them a significant advantage.

The majority of UAV-related research focuses on their use as a network node. Hadiwardoyo et al.[12] evaluated the impact of land topology on UAV-Vehicle communication and exhibited remarkable results in long-distance communication, reaching a distance of over 1km.

There are other examples of UAVs being employed in WSN (Wireless Sensors Network), as shown in the work of Zhan et al. [33], who presented an energy-efficient data collecting system based on a UAV-enabled WSN. A UAV can be deployed to quickly acquire data from a specific source thanks to its mobility.

When data is collected from a variety of sensors and sources, it is also necessary to examine how they will be merged.

Some studies have taken an ontology-approach to this solution, as shown by the review work of Bendadouche et al. [3]. Calder et al.[5], for instance, employed a reasoning technique to assess the behavior of multiple sensors in a coastal ecosystem. They used logical criteria to try to figure out if a sensor was working properly: Did the sensor record a reading? Was it completed at the appropriate time? Is the recorded value within a reasonable range?

Compton et al.[8] also created a sensor-specific ontology. They intended to build a model with enough abstraction to allow them to do so. As a result of this work, it would be simple to add new sensors as well as read the data collected. This work, along with Calder et al. [5], was eventually merged into the SNN (Semantic Sensor Network) ontology [17], which is described as an *"ontology for describing sensors and their observations, the involved procedures, the studied features of interest, the samples used to do so, and the observed properties, as well as actuators."*

From this literature review, the following conclusions can be drawn:

- Inclement weather, particularly rain, reduces visibility and is a major cause of car accidents.
- In an increasingly connected environment, intelligent tools such as UAVs can be requested to offer more data to increase perception and sight.
- When it comes to multi-sensor management, knowledge bases, particularly ontologies, have shown considerable promise in prior studies.

These deductions are the foundation for the methods described in this approach.

3 Ontology

Using various sources of data necessitates a way to federate and manage them. There are several options, one of them is the use of a knowledge base [27]. A knowledge-based model can be implemented in a variety of methods, including logic programming [16], knowledge graphs [23], and expert systems[2].

An ontology is another interesting way of doing so. Ontology is defined as "a formal explicit description of ideas in a domain of discourse, characteristics of each concept specifying various traits and attributes of the concept, and constraints on slots" by the Stanford 101 Guide[21]. The key actors within a realm of discourses, as well as the various interactions and links between them, are represented in an ontology. The following are some of the important elements of an ontology:

- Classes: Describe the concepts in the domain, whether they are abstract ideas or physical actors. Levels can be used to organize classes; for example, a Vehicle can be the top-level class, with Car, Bus, and Bike as sub-classes.

- Properties: The information about classes that is exclusive to them. They might be inherent to an object or extrinsic, expressing the links between several concepts.
- Individuals: Real instances representing the elements of the ontology.

Reasoners can be used to add a layer of intelligence and reflection to a knowledge base, in addition to being able to represent all of the acting elements. A reasoner is a tool that can derive logical conclusions from a set of facts, making it easier to classify the elements of an ontology. For example, if an instance V is declared as a Car, and the class Car is a sub-class of vehicle, then the reasoner infers that V is a vehicle [14].

Some reasoners can be provided by SWRL rules (Semantic Web Rule Language) [22] in a more complex situation. It is a logic description language that allows to combine different rules to create a more sophisticated axiom.

This study's ontology concerns both the vehicle and its surroundings. There are a lot of interconnected classes, but only a few of them make up the application's core:

- **Vehicle** representing the various vehicles that have been identified in the surrounding area. Both the Car and UAV entities are included in this class.
- **Weather** is a list of all the different types of weather that can occur. In this case *[Sunny,Fog,Rain,Snow]*.
- **Environment** describes the setting in which the vehicle evolves, one amongst *[NormalEnv,DarkEnv,BadWeatherEnv,UnusualEnv]*.
- **Sensors** refers to the sensors that are utilized in a vehicle's perception system, as detailed in [29]. The main ones are *[cameraMono,cameraStereo,cameraInfra, Lidar,Radar,Sonar]*. In addition, there are also environmental sensors used to determine the environment status, *[rainSensor,brightnessSensor,fogSensor]*.

In addition to the population of the knowledge base, logical rules are used to offer an extra layer of intelligence.

The acquired environmental data are continuously logged by the environmental sensors. The reasoner engine infers a new Environment class when a specific threshold is reached, and the relevant Sensors are triggered accordingly.

4 Use Case and Testing

The model was tested and validated in a simulated scenario. The Udacity project [28], a car simulator based on the Unity engine [11], was used to create the interface. It enables for the creation of driving environments (roads, barriers) and driving circumstances (rain, fog, physics limits, etc.), as well as manual and automatic control of the car.

Technically, the driving data is saved in JSON format and delivered to a Python engine, which parse and read the data in order to populate the ontology. The reasoner will then be summoned to determine the state of the environment. It will also determine which UAVs and sensors should be turned on, as well as write the result data in a file that the simulator can access and read.

The inferring of the correct weather and the management of sensors, a set of logical rules has been implemented. Here is an example of a set of rules for the rain detection:

```
1. RainSensor(?rainS) ^ hasRainValue(?rainS, ?rainV)^
        swrlb:greaterThan(?rainV, 50) ^Weather(?W)
                ->Rain(?W)

2. Rain(?R) ^ Environment(?Env)
                -> BadWeather(?Env)

3. Rain(?R)^ UAV(?u)^ Radar(?ra)^ cameraInfra(?c)^
        hasSensor(?u,?c)^ hasSensor(?u,?ra)
                -> isActiveUAV(?u,true)
```

Once the ontology is populated with the data gathered from the simulator and the data properties filled, the data sent by the sensor is compared to a threshold. If the value is greater than the threshold, then the system knows that the vehicle is in a rainy situation, and infers the environment as such.

The system will then check if there is a UAV nearby. If so, it will then ensure that the embedded sensors allow the detection in a rainy situation (As stated before, some sensors do not function properly in a bad weather). If applicable, the UAV will then transmit the data it gathers to the vehicle.

4.1 Simulation Tests

In order to validate the model, the following scenario is considered: *The controlled vehicle moves on a road blocked by an immobile car. The vehicle's sensors try to detect the obstacle in a clear and foggy situations.*

As shown in Fig. 1, in the first case, the weather is clear and the visibility is good. A UAV is present, but according to the logical rules, there is no need to ask for its support.

However, the second scenario shown in Fig. 2 is more interesting. In the first case, there is no UAV available, and the obstacle is detected when the separating distance is really close only.

In the third case, in Fig. 3, a UAV is present. After ensuring that the sensors on it are efficient in a Foggy situation, perception data are requested, and the obstacle is detected sooner. In this situation, the simulation data are sent to the knowledge base and it can identify the correct weather environment. The gathered data are also transmitted to the vehicle, which can detect potential obstacles on the road. Figure 4 shows that the reasoner has detected that the environment is foggy. If that is the case, some sensors are needed to aide in the perception of the driving environment.

Fig. 1 Simulator with no fog and no UAV assist

Fig. 2 Simulator with fog and no UAV assist

Fig. 3 Simulator with fog and UAV assist

Fig. 4 Python widget displaying the results inferred by the ontology

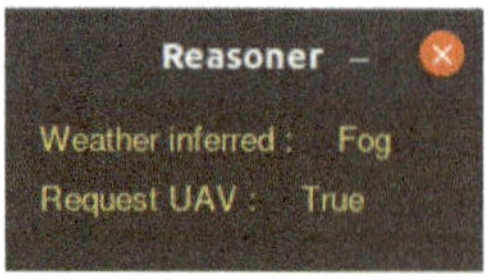

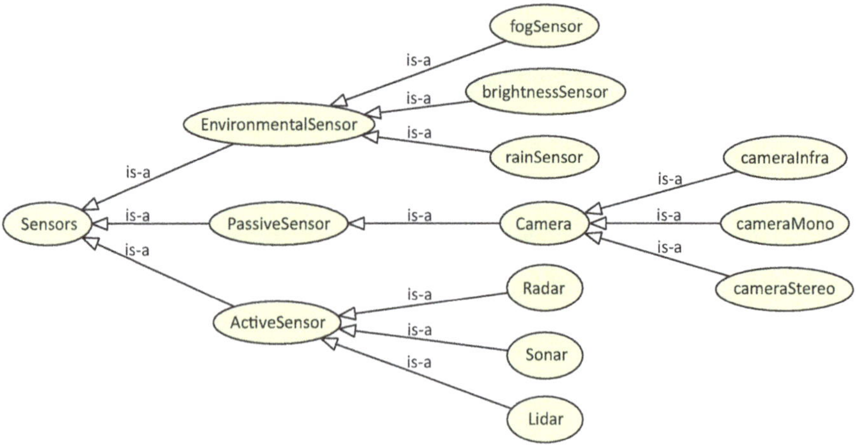

Fig. 5 Illustration of the "Sensors" class

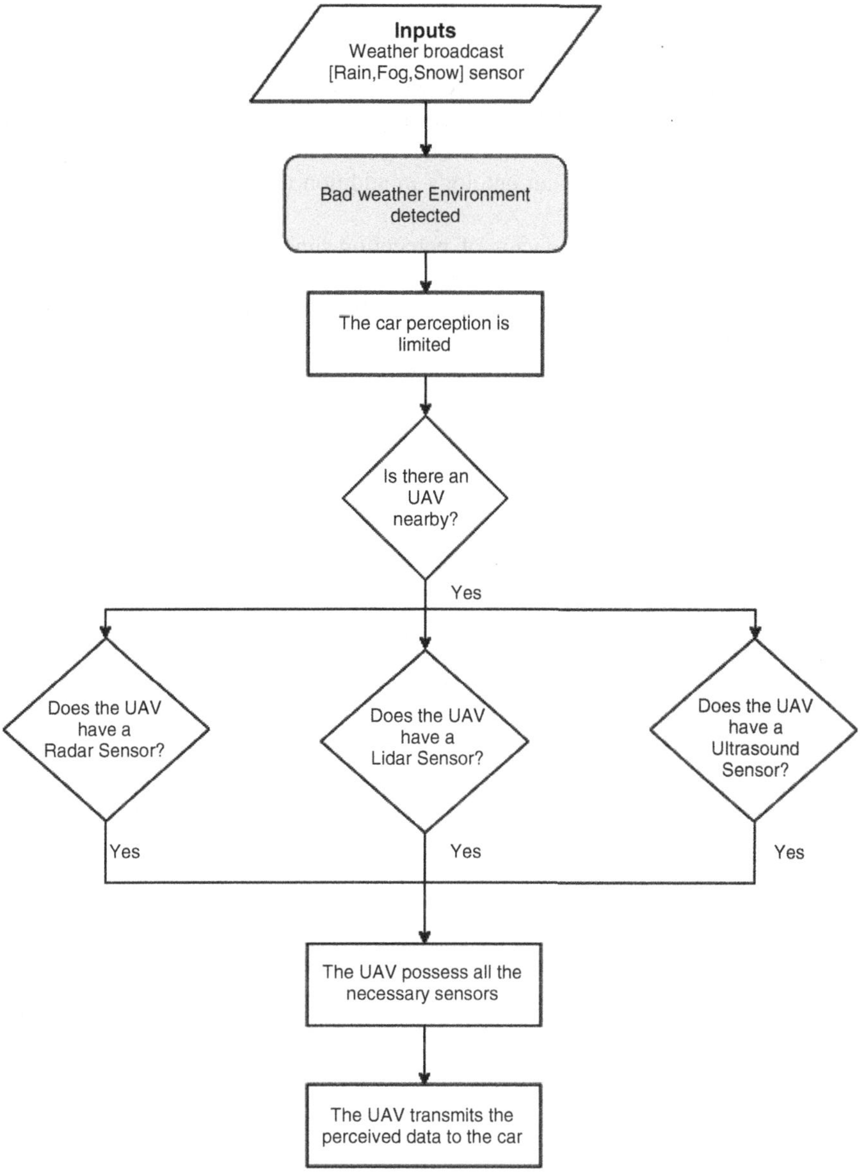

Fig. 6 Flowchart illustrating the interaction process

Figure 5 shows the ontology of the Sensors class. Figure 6 shows the algorithm that is used when bad environment (including foggy weather) is detected. In such a case, when a UAV is detected and it has any of the following sensors: radar, lidar and ultrasound sensor, then such sensor is activated to assist in the perception of the driving environment.

5 Conclusion

This work described the use of a knowledge base for vehicular perception enhancement in a Drone/Vehicle collaborative environment. The embedded sensors of the drone gather information on the surroundings and transmit them to the vehicle, where they will be stored in an ontology in addition to the data already gathered by the vehicle.

Bad weather is a known factor of perception hindrance, and an autonomous vehicle might end up lacking the correct amount of information in order to take a decision. This could be solved by having a drone gather additional information and transmit them to the vehicle. The two sources of data would be merged in a knowledge base and the incoming information are labeled and classified according to their nature. In addition to the knowledge representation, a layer of intelligence is added to the model thanks to the use of logical rules: The different elements of the knowledge base are submitted to a reasoner which can identify the correct state of the environment, as well as request the drone assistance if it is deemed adequate and take the correct decisions. As a result, the vehicle may improve its awareness and detect obstructions in adverse weather by using the drone's sensors. The model proposed in this work was tested and validated in a simulated environment.

This study will continue by considering multiple axes of improvement:

- **Consideration of other environments:** in addition to weather, there are other variable which can alter a sensor's operation, such as brightness or temperature. It would be interesting to add environment situations depending of those parameters
- **Communication protocol:** there are multiple ways for connected entities to communicate, each having strengths and weaknesses. For example, the use of VLC (Visible Light Communication) could be interesting in some specific environments, especially if coupled with another communication protocol
- **Simulator improvement:** the simulator already generates an important amount of data of different types, and this study only made use of a part of them.

References

1. Andrey, J., Yagar, S.: A temporal analysis of rain-related crash risk. Accid. Anal. Prev. **25**(4), 465–472 (1993)
2. Balci, O., Smith, E.P.: Validation of expert system performance. Technical report. Department of Computer Science, Virginia Polytechnic Institute & State ... (1986)
3. Bendadouche, R., Roussey, C., de Sousa, G., Chanet, J.P., Hou, K.M.: Etat de l'art sur les ontologies de capteurs pour une intégration intelligente des données. 30e Congrès INFORSID (Informatique des Organisations et Systèmes d'Information et de Décision) 2012, Montpellier, France, p. 16. (29-31 May 2012)

4. Bengler, K., Dietmayer, K., Farber, B., Maurer, M., Stiller, C., Winner, H.: Three decades of driver assistance systems: review and future perspectives. IEEE Intell. Transp. Syst. Mag. **6**(4), 6–22 (2014). https://doi.org/10.1109/MITS.2014.2336271, http://ieeexplore.ieee.org/document/6936444/
5. Calder, M., Morris, R.A., Peri, F.: Machine reasoning about anomalous sensor data. Eco. Inform. **5**(1), 9–18 (2010). https://doi.org/10.1016/j.ecoinf.2009.08.007, https://linkinghub.elsevier.com/retrieve/pii/S1574954109000715
6. Campbell, M.: The wet-pavement accident problem: breaking through. Traffic Q. **25**(2) (1971)
7. Campbell, M., Egerstedt, M., How, J.P., Murray, R.M.: Autonomous driving in urban environments: approaches, lessons and challenges. Phil. Trans. R. Soc. A. **368**(1928), 4649–4672 (2010). https://doi.org/10.1098/rsta.2010.0110, https://royalsocietypublishing.org/doi/10.1098/rsta.2010.0110
8. Compton, M., Neuhaus, H., Taylor, K., Tran, K.N.: Reasoning about sensors and compositions. 2009 International Workshop on Semantic Sensor Networks. (26 October 2009), Washington, DC, USA, pp. 33–48
9. Das, S., Brimley, B.K., Lindheimer, T.E., Zupancich, M.: Association of reduced visibility with crash outcomes. IATSS Res. **42**(3), 143–151 (2018). https://doi.org/10.1016/j.iatssr.2017.10.003, https://linkinghub.elsevier.com/retrieve/pii/S0386111216300681
10. Das, S., Dutta, A., Sun, X.: Patterns of rainy weather crashes: applying rules mining. Journal of Transportation Safety and Security. **12**(9), 1083–1105 (2020)
11. Haas, J.K.: A History of the Unity Game Engine (2014)
12. Hadiwardoyo, S.A., Hernández-Orallo, E., Calafate, C.T., Cano, J.C., Manzoni, P.: Experimental characterization of UAV-to-car communications. Elsevier Computer Networks. **136**, 105–118 (8 May 2018)
13. Harith, S.H., Mahmud, N., Doulatabadi, M.: Environmental factor and road accident: a review paper. IOEM 9th International Conference on Industrial Engineering and Operations Management Bangkok, Thailand, p. 10, (5-7 March 2019)
14. Hina, M.D., Thierry, C., Soukane, A., Ramdane-Cherif, A.: Cognition of driving context for driving assistance. World Acad. Sci. Eng. Technol. Int. J. Comput. Inf. Eng. **12**(2), 11 (2018)
15. Iwasaki, Y.: A method of robust moving vehicle detection for bad weather using an infrared thermography camera. In: International Conference on Wavelet Analysis and Pattern Recognition, p. 5 (2008)
16. Jaffar, J., Maher, M.J.: Constraint logic programming: a survey. J. Log. Program. **19–20**, 503–581 (1994). https://doi.org/10.1016/0743-1066(94)90033-7
17. Lefort, L., Henson, C., Taylor, K., Barnaghi, P., Compton, M., Corcho, O., García Castro, R., Graybeal, J., Herzog, A., Janowicz, K., et al.: Semantic sensor network XG final report (2011)
18. Mahlisch, M., Schweiger, R., Ritter, W., Dietmayer, K.: Sensorfusion using spatio-temporal aligned video and lidar for improved vehicle detection. In: 2006 IEEE Intelligent Vehicles Symposium, pp. 424–429. IEEE, Meguro-Ku (2006). https://doi.org/10.1109/IVS.2006.1689665, http://ieeexplore.ieee.org/document/1689665/
19. Menouar, H., Guvenc, I., Akkaya, K., Uluagac, A.S., Kadri, A., Tuncer, A.: UAV-enabled intelligent transportation systems for the smart city: applications and challenges. IEEE Commun. Mag. (2017). https://doi.org/10.1109/MCOM.2017.1600238CM
20. New Jersey bill a2757. Senate and General Assembly of the State of New Jersey (2012)
21. Noy, N., Mcguinness, D.: Ontology development 101: a guide to creating your first ontology. Knowl. Syst. Lab. **32** (2001)
22. O'Connor, M., Nyulas, C., Shankar, R., Das, A., Musen, M.: The SWRLAPI: A Development Environment for Working with SWRL Rules, p. 5
23. Paulheim, H.: Knowledge graph refinement: a survey of approaches and evaluation methods. Semant. Web **8**(3), 489–508 (2016). https://doi.org/10.3233/SW-160218, https://www.medra.org/servlet/aliasResolver?alias=iospress&doi=10.3233/SW-160218
24. Rasshofer, R.H., Gresser, K.: Advances in Radio Science, p. 5
25. Smith, M.: Light Detection and Ranging (LIDAR), vol. 2. A bibliography with abstracts. NTIS (1978)

26. Su, K., Li, J., Fu, H.: Smart city and the applications. In: 2011 International Conference on Electronics, Communications and Control (ICECC), pp. 1028–1031. IEEE, Piscataway (2011)
27. Trochim, W.M., Donnelly, J.P.: Research Methods Knowledge Base, vol. 2. Atomic Dog Pub., Mason (2001)
28. Udacity: Udacity self-driving car project (2017). https://github.com/udacity/self-driving-car-sim
29. Van Brummelen, J., O'Brien, M., Gruyer, D., Najjaran, H.: Autonomous vehicle perception: the technology of today and tomorrow. Transp. Res. C Emerg. Technol. **89**, 384–406 (2018). https://doi.org/10.1016/j.trc.2018.02.012, https://linkinghub.elsevier.com/retrieve/pii/S0968090X18302134
30. Vanholme, B., Gruyer, D., Lusetti, B., Glaser, S., Mammar, S.: Highly automated driving on highways based on legal safety. IEEE Trans. Intell. Transp. Syst. **14**(1), 15 (2013)
31. Woll, J.: Monopulse Doppler radar for vehicle applications. In: Proceedings of the Intelligent Vehicles '95. Symposium, pp. 42–47. IEEE, Detroit (1995). https://doi.org/10.1109/IVS.1995.528255, http://ieeexplore.ieee.org/document/528255/
32. World Health Organization: Save LIVES: A Road Safety Technical Package. World Health Organization, Geneva (2017). oCLC: 1131688778
33. Zhan, C., Zeng, Y., Zhang, R.: Energy-efficient data collection in UAV enabled wireless sensor network. arXiv:1708.00221 [cs, math] (2017). http://arxiv.org/abs/1708.00221

Adaptive Backstepping-RBF Control of Lower Limb Exoskeleton

Farid Kenas, Noureddine Ababou, and Nadia Saadia

1 Introduction

Recently, in the field of assistance to a person with motor disorders, many applied researches are oriented on the development of mechanized devices for assistance to the locomotion or functional rehabilitation of the lower limbs. In fact, the applications aim to restore original motor functions as well as to increase physical capacities in terms of muscle strength, speed of execution, and resistance to external constraints. Currently, in this field, many research laboratories in electronic instrumentation, biomechanics of human gesture, and medical robotics are involved in motorized exoskeletons, intelligent prostheses, and reduced gravity conveyors [1–3].

In reduced gravity, the automated devices for walking aim to relieve part of the body weight and contribute to the functional rehabilitation of a subject during his convalescence. In fact, they are mainly composed of a harness fixed to gallows surmounting a motorized treadmill [4, 5]. Besides, the control of such exoskeletons is performed using the information delivered by MEMS biosensors and inertial

F. Kenas (✉)
Laboratory of Instrumentation, University of Science and Technology Houari Boumediene Algiers, Bab-Ezzouar, Algeria

Laboratory of Robotics Parallelism and Embedded Systems, University of Science and Technology Houari Boumediene Algiers, Bab-Ezzouar, Algeria

N. Ababou
Laboratory of Instrumentation, University of Science and Technology Houari Boumediene Algiers, Bab-Ezzouar, Algeria

N. Saadia
Laboratory of Robotics Parallelism and Embedded Systems, University of Science and Technology Houari Boumediene Algiers, Bab-Ezzouar, Algeria

M. D. Hina et al. (eds.), *Advances in Computational Intelligence and Communication*, EAI/Springer Innovations in Communication and Computing,
https://doi.org/10.1007/978-3-031-19523-5_2

sensors [6]. Moreover, in the literature, several types of exoskeleton control are proposed, and these controllers are based on human intelligence, such as controllers based on neural networks [7, 8], sliding mode [9], backstepping [10], PID [11], neuron-fuzzy [12], and model-free control [7, 13, 14].

It is important to note that the identification phase of the unknown parameters of the dynamic model of our system (usually exoskeleton + body) is the crucial step before the design of the exoskeleton controller.

It is very important to know that the dynamic model of the system is complex to study. In fact, this system contains the body linked with the exoskeleton. Unfortunately, several uncertainties are incorporated during the operating process, such as disturbance, noise, and imperfection. Indeed, the previously identified dynamic model precision and the accuracy of the exoskeleton controller are affected by these uncertainties, especially during the movement. Many studies are focusing on the development of new strategies and adaptive controllers to estimate these unknown uncertainties. In [15], the authors presented an RBFN-based adaptive backstepping sliding mode command to control an upper limb exoskeleton. In this study, an RBF network is used to estimate a term in the dynamic model, which is exposed to dynamic uncertainties, external disturbances, and modeling errors. In another hand, the authors in [16] proposed a control for lower-limb exoskeleton based on sliding modes and used an RBF network to estimate a term present in the control law. The latter groups uncertainties of the exoskeleton dynamics model and some elements of the sliding surface and its derivative.

In order to identify all the unknown parameters of the dynamic model (inertia, resistive torque, viscous friction, etc.), the present paper introduces several identification parameters in the design steps of the controllers. It is important to know that the identification phase and the calculation of the torques required for the exoskeleton control are performed at the same time.

First of all, three adaptive controllers based on the backstepping technique and first- and second-order sliding modes are proposed. These controllers are used to control an exoskeleton and estimate all unknown parameters of the dynamic model simultaneously.

After that, to estimate all unknown parameters of the dynamic model and provide the necessary torque, two neural networks with radial basis functions (RBF) with the backstepping technique are incorporated.

Throughout this study, four controllers are designed to control flexion/extension of the right lower limb joints (hip, knee, and ankle) of an exoskeleton virtual prototype using three software, which are SolidWorks, SimMechanics and MATLAB/ Robotics Toolbox.

2 Exoskeleton Model

The developed model of this work is shown in Fig. 1. It presents the right leg lower limb exoskeleton. This model is designed in order to lead lower limbs users fixed by straps to move during rehabilitation training.

The exoskeleton is attached to a harness, which is in, turn, linked to an external frame as illustrated in Fig. 1a. The dimensions of this later are given in Fig. 1b. There are three actuated DOFs (flexion/extension) at the hip, knee, and ankle. The material used for the mechanical part is Aluminum 6061 T6 with a typical yield strength of 275 MPa leading to a weight of the exoskeleton without a harness of 4.93 kg.

Since dynamic model of the exoskeleton can be taken as a combination of three pendulums. And the exoskeleton is attached to the harness, there is no downforce on the ground. Moreover, the reaction force between the exoskeleton and the ground is null.

Taking into consideration references [9, 11], the dynamic model of the exoskeleton can be done as:

$$\ddot{\theta}_j = \frac{1}{I_j}\left(\tau_j - fs_j \cdot \text{sign}\left(\dot{\theta}_j\right) - fv_j \cdot \dot{\theta}_j - m_j \cdot l_j \cdot g \cdot \sin\theta_j\right) \tag{1}$$

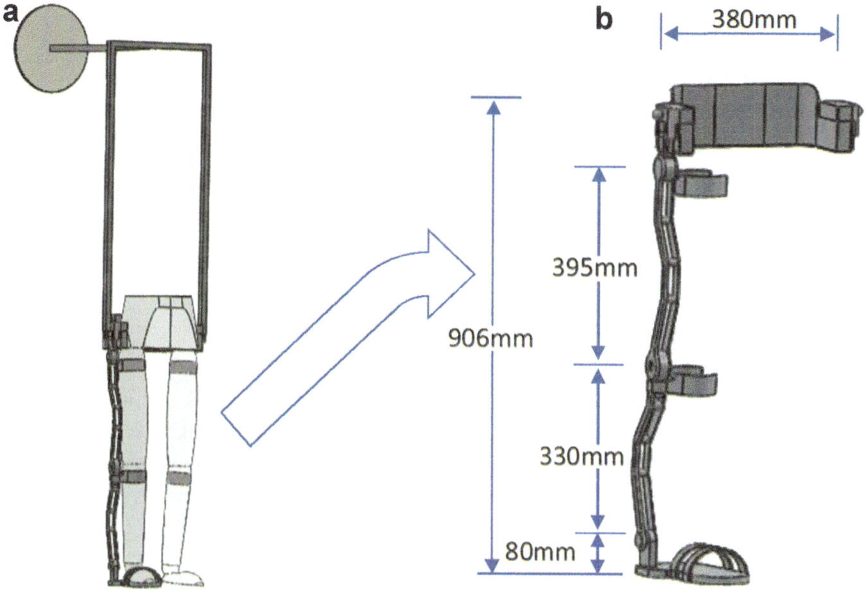

Fig. 1 (**a**) Exoskeleton attached to the harness (left) and (**b**) mechanical configuration in SolidWorks (right)

where θ_j, $\dot{\theta}_j$, and $\ddot{\theta}_j$ denote the angular positions, angular velocities, and angular accelerations of the exoskeleton joints, respectively. $j = (1, 2, 3)$ with $j = 1$ corresponds to the hip. $j = 2$ denotes the knee, and $j = 3$ corresponds to the ankle. $fs_j \cdot \text{sign}\left(\dot{\theta}_j\right)$ is the resistive torque due to dry friction. $fv_j \cdot \dot{\theta}_j$ is the resistive torque due to viscous friction, m_j is the mass of segment, $I_j$is Inertia, g *is* gravity, and l_j *is* length of segment.

The dynamic model of Eq. 1 can be expressed as a state space equation described by:

$$\begin{cases} \dot{x_1} = x_2 \\ \dot{x_2} = \frac{U}{I} + \frac{1}{I}\left(-fs \cdot \text{sign}\left(x_2\right) - fv \cdot x_2 - m \cdot l \cdot g \cdot \sin x_1\right) \end{cases} \tag{2}$$

where

$$\begin{cases} \theta = x_1 \\ \dot{\theta} = x_2 \\ \tau = U \end{cases} \tag{3}$$

This exoskeleton dynamic model shows unknown parameters of the system. The aim of this work is to find a control approach able to provide relatively high dynamic precision with the presence of unknown parameters in the dynamic model.

3 Controller Design

In order to ensure the convergence of the position θ toward the desired trajectory q_d, four different control designs are explored: backstepping techniques, first- and second-order sliding mode, and neural network.

3.1 Backstepping Control Design

The backstepping technique aims to calculate in several steps a command that guarantees the overall stability of the system and to identify the unknown parameters of a given system. It is important to note that the stability of this technique is based on the use of the Lyapunov function, which is positive definite and its derivative is negative [17].

The parameters to be estimated are done as in Eq. 4:

$$\begin{cases} \theta_1 = fv \\ \theta_2 = fs \\ \theta_3 = m \cdot l \cdot g \\ \beta = \frac{1}{I} \end{cases} \tag{4}$$

So, Eq. 2 becomes as given in Eq. 5:

$$\begin{cases} \dot{x_1} = x_2 \\ \dot{x_2} = \beta.U + \beta.(-\theta_1 \cdot \text{sign}(x_2) - \theta_2 \cdot x_2 - \theta_3 \cdot \sin x_1) \end{cases} \tag{5}$$

The tracking error and the velocity tracking error are defined as in Eq. 6.

$$\begin{cases} z_1 = x_1 - q \\ z_2 = x_2 - \alpha - \dot{q} \end{cases} \tag{6}$$

where q is the desired trajectory and α is the virtual control.

To guarantee the convergence stability of the system, the Lyapunov function $V(z_1)$ is chosen. This later must be positive definite around the state variable and can be written as in Eq. 7.

$$V_1 = \frac{1}{2} z_1^2 \tag{7}$$

The derivate of this function is shown as follows (Eq. 8):

$$\dot{V_1} = z_1 \cdot z_2 - c_1 \cdot z_1^2 \tag{8}$$

with $\alpha = -c_1 \cdot z_1$ and $c_1 > 0$.

The Lyapunov function $V(z_2)$ is related to the position, velocity tracking error, and estimation of unknown parameters and can be formulated as given in Eq. 9:

$$V_2 = V_1 + \frac{1}{2} z_2^2 + \frac{1}{2} \tilde{\theta}_i^T \Gamma^{-1} \tilde{\theta}_i + \frac{1}{2} \tilde{\beta}^T \gamma^{-1} \tilde{\beta} \tag{9}$$

with $\tilde{\theta}_i = \hat{\theta}_i - \theta_i, \tilde{\beta} = \hat{\beta} - \beta$,

$\tilde{\beta}, \tilde{\theta}_i$ are the estimations errors, and constants $\Gamma > 0, \gamma > 0$.
Whose derivative is:

$$\dot{V_2} = \dot{V_1} + z_2 \cdot \dot{z_2} + \tilde{\theta}_i^T \Gamma^{-1} \dot{\hat{\theta}}_i + \tilde{\beta}^T \gamma^{-1} \dot{\hat{\beta}} \tag{10}$$

$$\begin{aligned} \dot{V_2} = & -c_1 \cdot z_1^2 + z_2 \left(z_1 + \hat{\beta} \cdot U + \hat{\beta} \left(-\hat{\theta_1} \cdot \text{sign}(x_2) - \hat{\theta_2} \cdot x_2 - \hat{\theta_3} \cdot \sin x_1 \right) - \dot{\alpha} - \ddot{q} \right) \\ & + \tilde{\theta}_1 \Gamma^{-1} \left(\dot{\hat{\theta}}_1 + \hat{\beta} \cdot \Gamma \cdot \text{sign}(x_2) . z_2 \right) + \tilde{\theta}_2 \Gamma^{-1} \left(\dot{\hat{\theta}}_2 + \hat{\beta} \cdot \Gamma \cdot x_2 \cdot z_2 \right) \\ & + \tilde{\theta}_3 \Gamma^{-1} \left(\dot{\hat{\theta}}_3 + \hat{\beta} \cdot \Gamma \cdot \cos(x_1) \cdot z_2 \right) + \tilde{\beta} \gamma^{-1} \left(\dot{\hat{\beta}} - \gamma z_2 \left(U - \hat{\theta}_1 . \text{sign}(x_2) - \hat{\theta_2} \cdot x_2 - \hat{\theta}_3 . \sin x_1 \right) \right. \\ & \left. + \tilde{\beta} \cdot z_2 \left(\tilde{\theta}_1 \cdot \text{sign}(x_2) + \tilde{\theta}_2 \cdot x_2 + \tilde{\theta}_3 \cdot \sin(x_1) \right) \right) \end{aligned} \tag{11}$$

To guarantee the convergence of the z_2 to zero, the $\dot{V}_2$ should be semi-negative definite. This can be satisfied by ensuring the relationship of Eq. 12:

$$\dot{V}_2 = -c_1 \cdot z_1^2 - c_2 \cdot z_2^2 \;\; ; c_1, c_2 > 0 \tag{12}$$

With the control law formed in the Eq. 13:

$$U = \frac{1}{\hat{\beta}}\left(-c_2 z_2 - z_1 - \hat{\beta}\left(-\hat{\theta}_1 \cdot \mathrm{sign}(x_2) - \hat{\theta}_2 \cdot x_2 - \hat{\theta}_3 . \sin(x_1)\right) - c_1 x_2 + c_1 \cdot \dot{q} + \ddot{q}\right) \tag{13}$$

And the updates laws of the unknown parameters are given as follows:

$$\begin{cases} \dot{\hat{\theta}}_1 = -\hat{\beta} \cdot \Gamma \cdot \mathrm{sign}(x_2) \cdot z_2 \\ \dot{\hat{\theta}}_2 = -\hat{\beta} \cdot \Gamma \cdot x_2 \cdot z_2 \\ \dot{\hat{\theta}}_3 = -\hat{\beta} \cdot \Gamma . \sin(x_1) \cdot z_2 \\ \hat{\beta} = \gamma z_2 \left(U - \hat{\theta}_1 \cdot \mathrm{sign}(x_2) - \hat{\theta}_2 \cdot x_2 - \hat{\theta}_3 \cdot \sin x_1\right) \end{cases} \tag{14}$$

3.2 Backstepping-First-Order Sliding Mode Control Design

The principle of sliding mode control is based on imposing a dynamic to the system so that it converges toward a certain surface called sliding surface. This latter represents the desired dynamic of the system [9]. The command law for sliding mode includes an equivalent control U_{eq} and a discontinuous control U_{dis}. The calculation procedure remains the same as that developed in the backstepping control except the integration of the sliding surface S instead of z_2.

The form of sliding surface is presented in Eq. 15 as:

$$S = \lambda z_1 + x_2 - \alpha - \dot{q} \;\; ; \lambda > 0 \tag{15}$$

$\dot{V}_2$ is chosen as follows:

$$\dot{V}_2 = -c_1 \cdot z_1^2 - c_2 \cdot S^2 ; c_1, c_2 > 0 \tag{16}$$

With the equivalent control law given in Eq. 17 as:

$$U_{eq} = \frac{1}{\hat{\beta}}\left(S(-c_2 - \lambda) + z_1\left(\lambda^2 - 1\right) - \lambda\alpha + \dot{\alpha} + \ddot{q} + \hat{\beta}\left(\hat{\theta}_1 \cdot \mathrm{sign}(x_2) + \hat{\theta}_2 \cdot x_2 + \hat{\theta}_3 \cdot \sin x_1\right)\right) \tag{17}$$

with $\dot{\alpha} = (\lambda - c_1) \cdot (\dot{x_1} - \dot{q})$

whereas the updated laws of the unknown parameters are in Eq. 18:

$$\begin{cases} \dot{\hat{\theta}}_1 = -\hat{\beta}\cdot \Gamma\cdot \operatorname{sign}(x_2)\cdot S \\ \dot{\hat{\theta}}_2 = -\hat{\beta}\cdot \Gamma\cdot x_2\cdot S \\ \dot{\hat{\theta}}_3 = -\hat{\beta}\cdot \Gamma\cdot \sin(x_1)\cdot S \\ \dot{\hat{\beta}} = \gamma S \left(U - \hat{\theta}_1\cdot \operatorname{sign}(x_2) - \hat{\theta}_2\cdot x_2 - \hat{\theta}_3\cdot \sin x_1\right) \end{cases} \tag{18}$$

Finally, the control law for backstepping-first-order sliding mode can be derived as in Eq. 19:

$$U = U_{\text{eq}} + U_{\text{dis}} = U_{\text{eq}} + U_{\text{sat}} = U_{\text{eq}} - \frac{1}{\hat{\beta}}\cdot K\cdot \operatorname{sat}(S) \quad ; K > 0 \tag{19}$$

3.3 Backstepping-Second-Order Sliding Mode Control Design

During the use of the control law based on a first-order sliding mode, a phenomenon called chattering is observed on the control signal. The major disadvantages of such a phenomenon are as follows: a significant deterioration of system performance, instability, and possible damage of the system actuator [9]. Fortunately, the control by second-order sliding mode can be an adequate solution to this problem. In fact, this technique can ensure high performance in terms of tracking process and robustness [18]. Considering the second-order sliding mode command, the equivalent control U_{eq} remains unchanged, and the discontinuous control U_{dis} takes other forms.

The robustness of the second-order sliding mode controller varies according to the used algorithm. In this work, the super twisting algorithm is used due to the high performance obtained during the control of a knee exoskeleton [9]. This latter is given by Eqs. 20 and 21 as follows:

$$U_{\text{dis}} = U_1 + U_2 \tag{20}$$

$$\begin{cases} \dot{U}_1 = -r_1 \operatorname{sign}(S) \\ U_2 = -r_2 |S|^{\frac{1}{2}} \operatorname{sign}(S) \end{cases}$$

with constants

$$r_1, r_2 > 0 \tag{21}$$

The final control law for backstepping-second-order sliding mode is done as in Eq. 22:

$$
\begin{aligned}
U &= U_{\text{eq}} + \frac{1}{\hat{\beta}} U_{\text{dis}} \\
&= U_{\text{eq}} + \frac{1}{\hat{\beta}} U_{\text{dis}} = \frac{1}{\hat{\beta}} \left(S\left(-c_2 - \lambda\right) + z_1 \left(\lambda^2 - 1\right) - \lambda\alpha + \dot{\alpha} + \ddot{q} + \hat{\beta} \left(\hat{\theta_1} \cdot \operatorname{sign}\left(x_2\right) \right.\right. \\
&\quad \left.\left. + \hat{\theta_2} \cdot x_2 + \hat{\theta_3} \cdot \sin x_1 \right) \right) + \frac{1}{\hat{\beta}} U_{\text{dis}}
\end{aligned} \tag{22}
$$

with $\dot{\alpha} = (\lambda - c_1) \cdot (\dot{x_1} - \dot{q})$

and

$$
\begin{cases}
U_{\text{dis}} = U_1 + U_2 \\
\dot{U}_1 = -r_1 \operatorname{sign}(S) \\
U_2 = -r_2 |S|^{\frac{1}{2}} \operatorname{sign}(S)
\end{cases} \tag{23}
$$

And the update laws of the unknown parameters are done as in Eq. 24:

$$
\begin{cases}
\dot{\hat{\theta}}_1 = -\hat{\beta} \cdot \Gamma \cdot \operatorname{sign}\left(x_2\right) \cdot S \\
\dot{\hat{\theta}}_2 = -\hat{\beta} \cdot \Gamma \cdot x_2 \cdot S \\
\dot{\hat{\theta}}_3 = -\hat{\beta} \cdot \Gamma \cdot \sin\left(x_1\right) \cdot S \\
\dot{\hat{\beta}} = \gamma \cdot S \cdot \left(U_{\text{eq}} - \hat{\theta_1} \cdot \operatorname{sign}\left(x_2\right) - \hat{\theta_2} \cdot x_2 - \hat{\theta_3} \cdot \sin x_1 \right)
\end{cases} \tag{24}
$$

3.4 Backstepping-Neural Network (RBF)

In this part, a control based on the backstepping technique with artificial neural network is proposed. In fact, two unknown and continuous nonlinear functions φ_1 and φ_2 are estimated by a neural network with radial basis functions (RBF). Neural network weights are adaptively adjusted using online learning algorithm derived from Lyapunov theory [19–21]. The structure of the RBF is shown in Fig. 2.

According to Fig. 2, the mathematical model corresponding to the connection between inputs/outputs can be derived as given in Eq. 22:

$$
y_l(k) = p_{0,l} + \sum_{i=1}^{m} p_{i,l} \cdot \phi^* \left(\|x(k) - C_i\| \right) = P_i^T \cdot \phi^*(r) \tag{25}
$$

With C_i are the centers of the radial basis functions, $x(k)$ is the input vector, m is the number of neurons in the hidden layer, $p_{0,\,l}$ is the polarization, and P^T is the vector of weights.

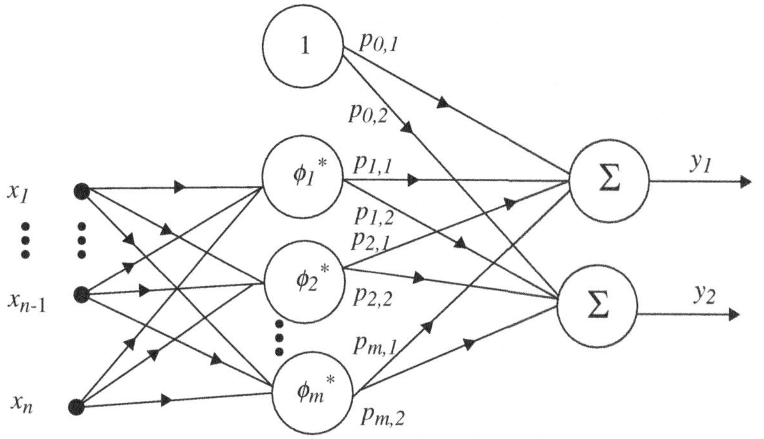

Fig. 2 Structure of RBF neural network with two outputs

The Gaussian activation function $\phi^*(r)$ is defined by Eq. 26 as follows:

$$\phi^*(r) = \exp\left(-\frac{r^2}{2\cdot\sigma^2}\right), \lambda > 0 \tag{26}$$

where σ is the standard deviation.

In order to calculate the control law and estimate the unknown parameters, the estimated functions (φ_1, φ_2) are considered as done in Eq. 27:

$$\begin{cases} \varphi_1 = \frac{1}{I} = \varphi_1^*\cdot w^T + \widetilde{\varphi}_1 \\ \varphi_2 = -fs\cdot \operatorname{sign}(x_2) - fv\cdot x_2 - m\cdot l\cdot g\cdot \sin x_1 = \varphi_2^*\cdot p^T + \widetilde{\varphi}_2\, a \end{cases} \tag{27}$$

where φ_1^* and φ_2^* are two functions that can be defined using the approximation by RBF networks.

The system Eq. (2) takes the following form:

$$\begin{cases} \dot{x_1} = x_2 \\ \dot{x_2} = \varphi_1\cdot U + \varphi_1\cdot\varphi_2 \end{cases} \tag{28}$$

First step is the same step in the design of the backstepping controller.

The second corresponding Lyapunov function is:

$$V_2 = V_1 + \frac{1}{2}z_2^2 + \frac{1}{2}\tilde{w}^T\gamma^{-1}\widetilde{w} + \frac{1}{2}\tilde{p}^T\Gamma^{-1}\widetilde{p} \tag{29}$$

with constant matrices $\Gamma = \Gamma^T > 0$, and $\gamma = \gamma^T > 0$; and estimations errors: $\widetilde{w} = w - \hat{w}$ and $\widetilde{p} = p - \hat{p}$.

whose derivative is formulated as in Eq. 30:

$$\dot{V}_2 = -c_1 \cdot z_1^2 + z_2 \left(z_1 + \varphi_1^* \cdot \hat{w} \cdot U + \varphi_1^* \cdot \hat{w} \cdot \varphi_2^* \cdot \hat{p} - \dot{\alpha} - \ddot{q}\right) + \tilde{w}^T \gamma^{-1} \left(\dot{\hat{w}} - \gamma . z_2 \cdot \varphi_1^* \cdot U - \gamma \cdot z_2 \cdot \varphi_1^* \cdot \varphi_2^* \cdot \hat{p}\right) + \tilde{p}^T \Gamma^{-1} \left(-\dot{\hat{p}} + \Gamma \cdot z_2 \cdot \varphi_2^* \cdot \varphi_1^* \cdot \hat{w}\right) \tag{30}$$

To guarantee the convergence of the z_2 to zero, the $\dot{V}_2$ should be semi-negative definite. This can be satisfied by choosing Eq. 31:

$$\dot{V}_2 = -c_1 \cdot z_1^2 - c_2 \cdot z_2^2 \;\; ; c_1, c_2 > 0 \tag{31}$$

The adaptation laws of neural networks weights are:

$$\begin{cases} \dot{\hat{w}} = \gamma \cdot z_2 \cdot \varphi_1^* \cdot \left(U + \varphi_2^* \cdot \hat{p}\right) \\ \dot{\hat{p}} = \Gamma \cdot z_2 \cdot \varphi_2^* \cdot \varphi_1^* \cdot \hat{w} \end{cases} \tag{32}$$

and the control law is done as

$$U = \frac{1}{\varphi_1^* \cdot \hat{w}} \left(-c_2 \cdot z_2 - z_1 - \varphi_1^* \cdot \hat{w} \cdot \varphi_2^* \cdot \hat{p} - c_1 \cdot x_2 + c_1 \cdot \dot{q} + \ddot{q}\right) \text{ with } c_{1,2} > 0 \tag{33}$$

3.5 Co-simulation Used Method

The SIMULINK/MATLAB platform is used to implement the previously designed and modeled system. In fact, the desired trajectory database was collected from OpenSim software [22], where ten cycles of healthy walking are considered. The co-simulation diagram of the system is shown in Fig. 3. Indeed, the controllers are implemented within the system, where a comparison study is fulfilled in terms of tracking errors and torques occurring during the operating process.

The BS.SMC is the Backstepping-sliding mode controller where the equivalent control U_{eq} is calculated via the backstepping controller in function of the sliding surface. U_{eq} will be added to the discontinuous control U_{dis} to calculate the final control law as indicated in red dashes in Fig. 3.

The NBSC is the neural backstepping controller where two unknown functions φ_1 and φ_2 are estimated by two neural networks (RBF) as indicated in blue dashes in Fig. 3.

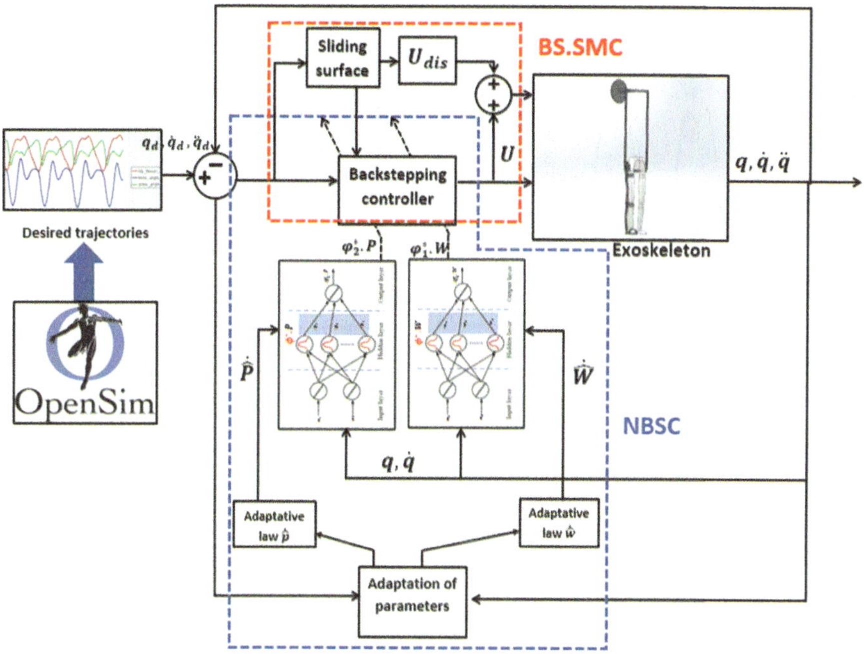

Fig. 3 Co-simulation diagram

4 Results and Discussions

In Fig. 4, both joint trajectories (hip, knee, and ankle) and the input torques for the backstepping controller are represented as a function of time.

It is clear that the black solid curves refer to the flexion-extension joint angles' desired values, and the red dashed ones present the measured angles at the actuators, whereas the blue solid curves correspond to the tracking errors.

Moreover, absolute mean error (AME), root mean square error (RMSE), standard deviation (SD), and maximal error (ME) are reported in Table 1.

In fact, analyzing the obtained results, it can be noted that the backstepping controller can easily follow the desired values for the three levels joints (hip, knee, and ankle). Besides, the RMSE accuracy during flexion or extension is less than 2°. But when changing the direction from extension to flexion or from flexion to extension, the backstepping controller fails to follow the desired values. In these cases, the maximum errors outrun 10° at the knee and ankle as illustrated on the left side of Fig. 4.

Through the evolution of the torques provided by the backstepping controller, it can be observed that several peaks appear. These peaks exceed 1 kN.m. Besides, the chattering phenomenon is showing in several times as exposed on the right side of Fig. 4.

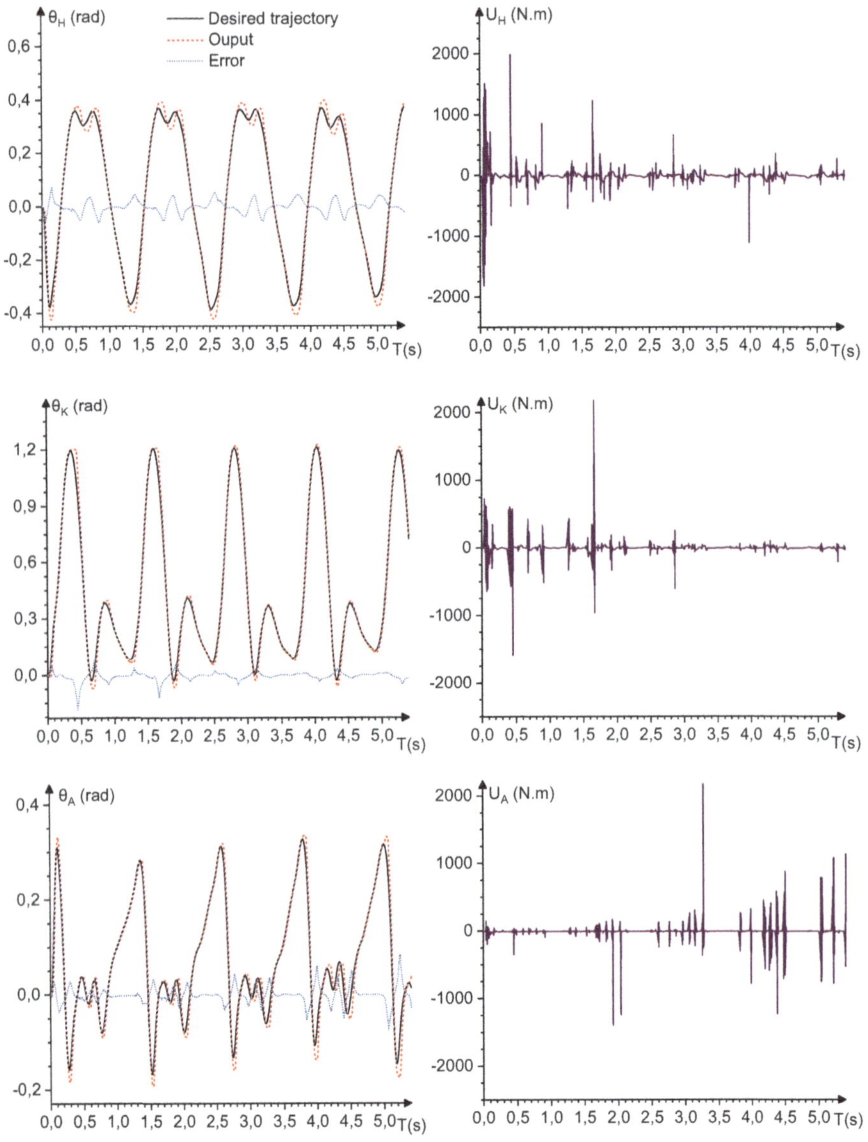

Fig. 4 Simulation results of hip, knee and ankle joints trajectories (θ_H, θ_K, θ_A) for backstepping controller and respective control input torques U_H, U_K, U_A

The use of the backstepping-first-order sliding mode controller gives slight improvements in RMSE and ME. Meanwhile, the use of the backstepping-second-order sliding mode controller gives less precise results than the backstepping controller. To address this problem, a radial-basis-function (RBF) neural network

Table 1 Absolute mean error, root mean square error, standard deviation, and maximal error for backstepping controller (BSC), backstepping-first-order sliding mode controller (BS first O. SMC), backstepping-second-order sliding mode controller (BS second O SMC) and neural backstepping controller (NBSC)

Joint	Controller	AME	RMSE	SD	ME
Hip	BSC	1,014°	1,373°	1,373°	4,330°
	BS 1st O. SMC	0,966°	1,308°	1,307°	4,303°
	BS 2nd O. SMC	1,020°	1,387°	1,386°	4,509°
	NBSC	0,298°	0,396°	0,390°	4,057°
Knee	BSC	0,838°	1,340°	1,332°	10,362°
	BS 1st O. SMC	0,879°	1,381°	1,372°	10,499°
	BS 2nd O. SMC	0,926°	1,481°	1,470°	10,364°
	NBSC	0,621°	0,769°	0,761°	2,669°
Ankle	BSC	1,031°	1,826°	1,825°	10,116°
	BS 1st O. SMC	0,950°	1,695°	1,694°	9,618°
	BS 2nd O. SMC	1,334°	2,412°	2,411°	12,047°
	NBSC	0,313°	0,470°	0,470°	2,132°

backstepping controller was used to minimize the tracking errors and to limit the chattering in the control torques.

The results obtained with the neural backstepping controller are shown in Fig. 5. It can be clearly noted that contrary to previous controllers, the use of the neural backstepping controller leads to torques without any peaks or chattering.

Analyzing the errors presented in Table 1, a great improvement is shown compared to the previously used controllers. Mainly, at the knee level, a maximum error of 2.6° is obtained instead of 10°. Whereas the RMSE is equal to 0.39°. Also, a great enhancement is observed at the ankle level with a maximum error of 2.1° instead of 10°. While the RMSE is only 0.47°.

Figure 6a summarizes the RMSEs obtained for each controller. It can be clearly noticed that the neural backstepping controller guarantees the best accuracy compared to the other controllers.

The RMSE in the case of the neural backstepping controller is (0,39°-0,76°-0,47°) for the ankle, knee, and hip, respectively, and for other controllers (backstepping, backstepping-first- and second-order sliding mode), the RMSE varies between 1.82° to 2.41° for the ankle, 1.34° to 1.48° for the knee, and 1.30° to 1.38° for the hip.

The use of the neural network with the backstepping controller leads to a reduction of the root mean square errors (RMSE) compared to backstepping controller and backstepping-first- and second-order sliding mode controllers where an accuracy improvement from 69% to 71% is observed at the hip and from 42% to 48% at the knee and from 72% to 80% at the ankle as shown in Fig. 6b, which summarizes the improvement rate in (%) for the RMSE brought by the neural backstepping controller compared to the other controllers.

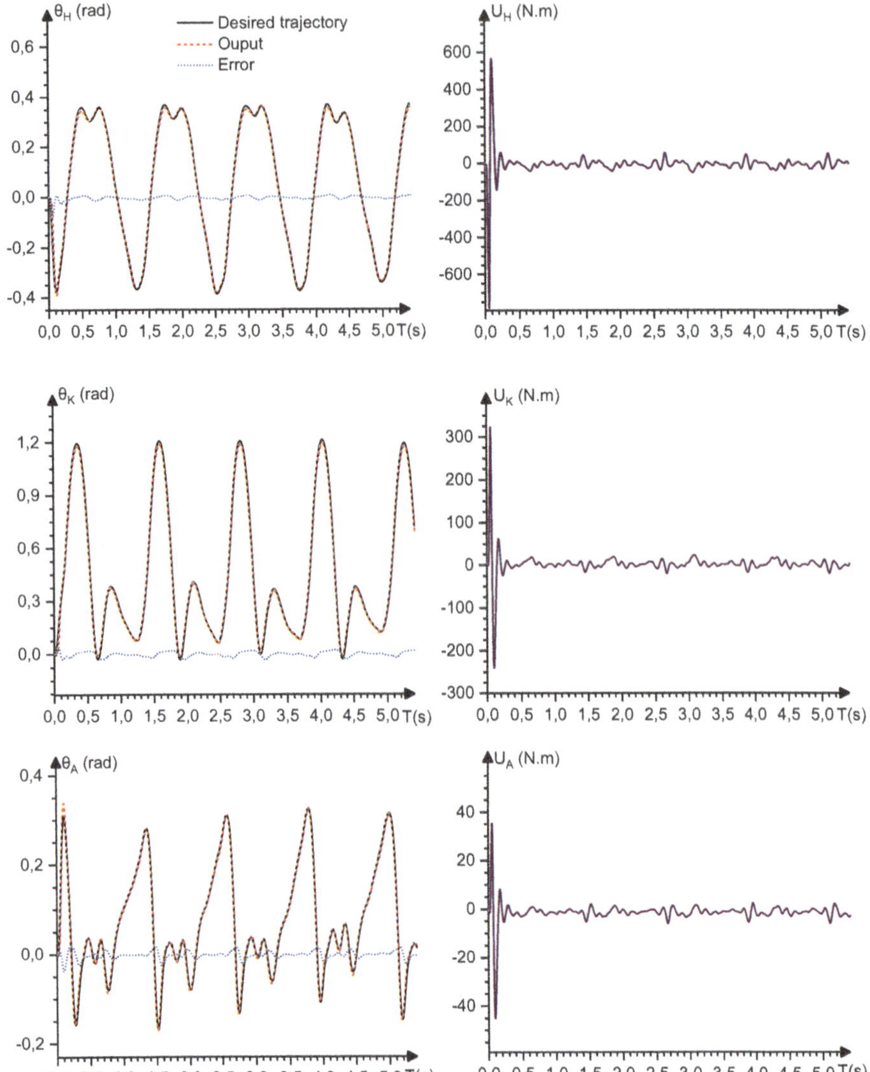

Fig. 5 Simulation results of hip, knee, and ankle joints trajectories (θ_H, θ_K, θ_A) for neural backstepping controller and respective control input torques U_H, U_K, U_A

5 Conclusion and Future Work

In this study, an adaptive neural backstepping (backstepping-RBF) was designed to control three joints (hip, knee, and ankle) of a right lower limb exoskeleton. The considered exoskeleton was built on the SolidWorks virtual prototype.

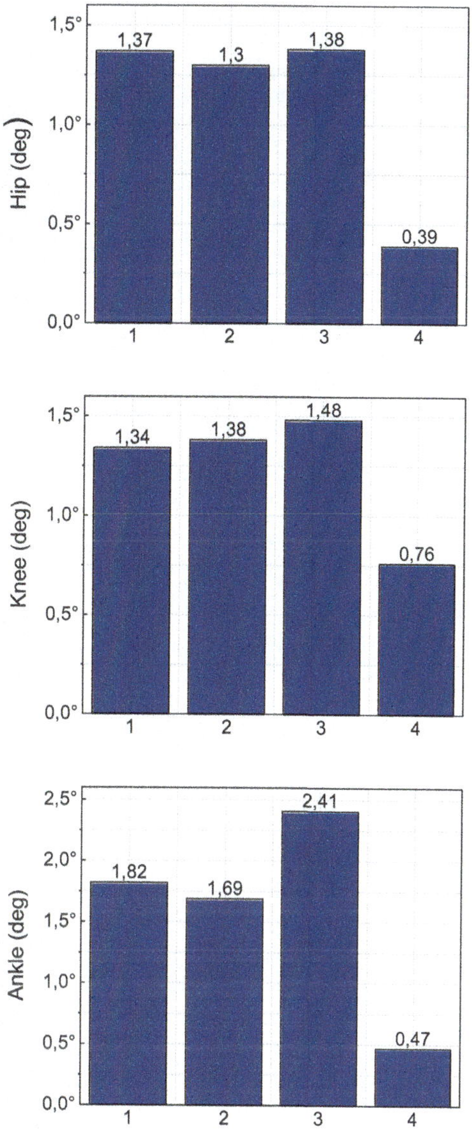

Fig. 6a RMS error for controllers ((1) backstepping, (2) backstepping-first-order sliding mode, (3) backstepping-second order sliding mode, (4) neural backstepping)

To demonstrate the performance of the proposed method, the systems of the co-simulation are implemented in MATLAB /SIMULINK and SolidWorks, respectively. The simulation results are compared to that obtained using the backstepping controller, backstepping-first- and second-order sliding mode controllers. The proposed method ensures the tracking of the desired trajectory with high accuracy while generating a torque without chattering.

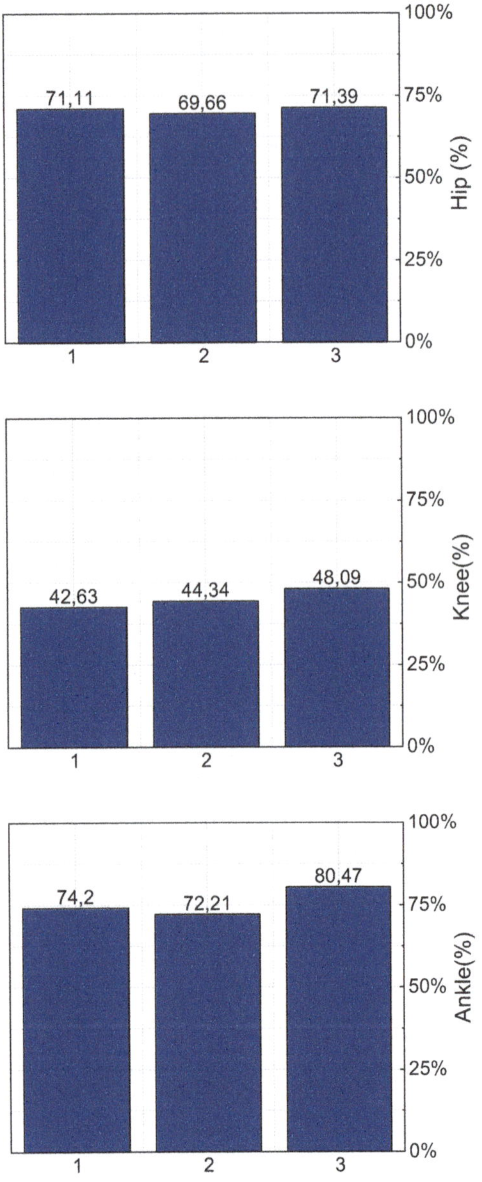

Fig. 6b Improvement rate in (%) of the RMS error brought by neural backstepping controller compared to ((1) backstepping, (2) backstepping-first-order sliding mode, (3) backstepping-second-order sliding mode)

The future works will focus on the improvement of the neural controller by using advanced controller techniques to choose the best parameters for RBF networks such as the k-mean algorithm, the use of other neural network types such as multilayer networks and varying the algorithms of the sliding modes in order to design a more accurate controller are also subjected. On the other hand, it is important to consider the extension of the exoskeleton mechanical structure by adding the left limb and increasing the number of DOF. For instance, adding abduction/adduction and internal/external rotation movements at the hip joint creates a complete exoskeleton for lower limbs.

References

1. Yang, F., Bhatt, T., Pai, Y.C.: Generalization of treadmill-slip training to prevent a fall following a sudden (novel) slip in over-ground walking. J. Biomech. **46**, 63–69 (2013)
2. Patil, S., Steklov, N., Bugbee, W.D., Goldberg, T., Colwell Jr., C.W., D'Lima, D.D.: Anti-gravity treadmills are effective in reducing knee forces. J. Orthop. Res. **31**(5), 672–679 (2013)
3. Mikami, Y., Fukuhara, K., Kawae, T., Kimura, H., Ochi, M.: The effect of anti-gravity treadmill training for prosthetic rehabilitation of a case with below-knee amputation. Prosthet. Orthot. Int. **39**, 502–506 (2015)
4. Hidayah, R., Jin, X., Chamarthy, S., Fitzgerald, M.M., Agrawal, S.K.: Comparing the performance of a cable-driven active leg exoskeleton (C-ALEX) over-ground and on a treadmill. In: 2018 7th IEEE International Conference on Biomedical Robotics and Biomechatronics (Biorob), pp. 299–304 (2018)
5. Goldberg, S.R., Stanhope, S.J.: Sensitivity of joint moments to changes in walking speed and body-weight-support are interdependent and vary across joints. J. Biomech. **46**(6), 1176–1183 (Apr. 2013)
6. Wheeler, J. et al.: "In-sole MEMS pressure sensing for a LowerExtremity exoskeleton," the first IEEE/RAS-EMBS international conference on biomedical robotics and biomechatronics, 2006. BioRob 2006, pp. 31–34
7. Zhang, X., Wang, H., Tian, Y., Peyrodie, L., Wang, X.: Model-free based neural network control with time-delay estimation for lower extremity exoskeleton. Neurocomputing. **272**, 178–188 (Jan. 2018)
8. Han, S., Wang, H., Tian, Y., Christov, N.: Time-delay estimation based computed torque control with robust adaptive RBF neural network compensator for a rehabilitation exoskeleton. ISA Trans. **97**, 171–181,2019 (2019)
9. SaberMefoued, "Commande robuste référencée intention d'une orthèse active pour l'assistance fonctionnelle aux mouvements du genou", PhD, University of Paris-Est, 2012
10. Lu, R., Li, Z., Su, C.-Y., Xue, A.: Development and learning control of a human limb with a rehabilitation exoskeleton. IEEE Trans. Ind. Electron. **61**(7), 3776–3785 (Jul. 2014)
11. Ajayi, M. O.: Modelling and control of actuated lower limb exoskeletons: a mathematical application using central pattern generators and nonlinear feedback control techniques. In: General Mathematics [math.GM], Université Paris-Est, 2016
12. He, H., Kiguchi, K.: A study on EMG-based control of exoskeleton robots for human lower-limb motion assist. In: Information Technology Applications in Biomedicine, 2007. ITAB 2007. 6th International Special Topic Conference on, IEEE, pp. 292–295 (2007)
13. Wang, H., Xu, H., Yang, T.: Hao tang, α-variable adaptive model free con-trol of iReHave upper-limb exoskeleton. Adv. Eng. Softw. **148** (2020)
14. P. Yang, J. Sun, J. Wang, G. Zhang and Y. Zhang, "Model-free based back-stepping sliding mode control for wearable exoskeletons," 2019 25th International Conference on Automation and Computing (ICAC), pp. 1–6, 2019
15. Wu, Q., et al.: RBFN-based adaptive Backstepping sliding mode control of an upper-limb exoskeleton with dynamic uncertainties. IEEE Access. **7**, 134635–134646 (2019)
16. Tu, Y., et al.: An adaptive sliding mode variable admittance control method for lower limb rehabilitation exoskeleton robot. Appl Sci. **10**(7), art. no. 2536 (2020)
17. Soukkou, Y.: Commande adaptative par backstepping d'une classe des systèmes non linéaires incertains, Magister Thesis, University of Jijel, 2014

18. Bartolini, G., Ferrara, A., Usani, E.: Chattering avoidance by second-order sliding mode control. IEEE Trans. Automat. Contr. **43**, 241–246 (1998)
19. Tran, D.T., Nguyen, M.N., Ahn, K.K.: RBF neural network based Backstepping control for an electrohydraulic elastic manipulator. Appl. Sci. **9**, 2237 (2019)
20. Liu, W.: Adaptive backstepping neural control for nonlinear systems with quantized input signals. In: 2017 Chinese Automation Congress (CAC), Jinan, China, pp. 1029–1034 (2017)
21. Liu, J.K.: Radial Basis Function (RBF) neural network control for mechanical systems, design, analysis and matlab simulation. Springer, New York (2013)
22. Delp, S.L., Anderson, F.C., Arnold, A.S., Loan, P., Habib, A., John, C.T., Guendelman, E., Thelen, D.G.: OpenSim: open-source software to create and analyze dynamic simulations of movement. I.E.E.E. Trans. Biomed. Eng. **54**(11), 1940–1950 (2007)

An Evaluation Framework for Blockchain-Based Public Key Infrastructures

Lukas Ehrmann, Kai-Oliver Kohlen, Feng Cheng, and Pejman Najafi

1 Introduction

As more and more services are moved to the cloud, the demand for secure identification and authentication is ever increasing. This problem has been around for almost as long as digital systems and the Internet exist. Different solutions and systems have been established to address the problem, such as the domain name system (DNS) and its security measures, accounts with credentials for users, or public-private key pairs. One of the most notable options is the X.509 Public Key Infrastructure (PKI). It is currently used to ensure authentication in various processes, such as software downloads, emails, and most importantly TLS connections.

This PKI is managed and maintained by the certificate authorities (CAs) as the core infrastructure. Due to the importance of the PKI for digital services, the CAs form a market worth over USD 1.7 billion, which is projected to grow significantly in the next years [29].

Given the size and importance of the current PKI, security against threats as well as reliable functioning became paramount for the provisioning of digital services. However, successful attacks in the last decade have shown weaknesses in the current PKI design: CAs like DigiNotar [2], Comodo [15], and Symantec [21] have been liable for issuing unauthorized certificates due to active attacks, technical or organizational failures. While these attacks were eventually discovered and the

Authors Lukas Ehrmann and Kai-Oliver Kohlen have contributed equally.

L. Ehrmann · K.-O. Kohlen (✉) · F. Cheng · P. Najafi
Hasso Plattner Institute, University of Potsdam, Potsdam, Germany
e-mail: Lukas.Ehrmann@student.hpi.de; Kai-Oliver.Kohlen@student.hpi.de; Feng.Cheng@hpi.de; Pejman.Najafi@hpi.de

M. D. Hina et al. (eds.), *Advances in Computational Intelligence and Communication*, EAI/Springer Innovations in Communication and Computing,
https://doi.org/10.1007/978-3-031-19523-5_3

certificates were revoked, they show that the compromise of just one CA threatens the security of all TLS-based connections. With dozens of CAs in existence, one therefore only needs to compromise the weakest link in the chain.

In addition, developments like the pervasiveness of IoT devices drastically increase the number of participants that need an authenticated identity. These changing requirements for a PKI in combination with the mentioned attacks inspired a far-reaching change in thinking of how a PKI is built. One emerging idea is to base a PKI on distributed ledger technology, which ensures the integrity and consistency of the certificates, as well as publishing them reliably. This idea has been explored since 2014 [19] and researchers created numerous specifications and implementations based on this idea.

However, the use of all these ideas remains unclear because there is no singular set of goals they want to meet and can be measured by. While Brunner et al. [9] compared different approaches, they focused on categorizing the papers based on technical decisions and performed evaluations. In this paper, we define essential requirements that blockchain-based PKIs must fulfill. Thereby, we understand essential requirements both with a technical dimension and an organizational, economic, user-oriented, and political focus. Requirements NIST [7] evaluated in 1994 included organizational, user-oriented, economic, and political aspects but are outdated. By combining, modifying, and extending the requirements of Brunner et al. and NIST, we enable the holistic and objective analysis of blockchain-based PKIs regarding their feasibility.

Our contributions to this field are threefold: we provide a holistic evaluation framework to analyze proposed solutions and their practicability. We reviewed and evaluated the existing solutions by using the proposed framework. Finally, we note unresolved challenges that need to be addressed in the future to allow the adoption of a blockchain-based PKI design.

The structure of this paper is as follows: In Sect. 2, we introduce the current state of the PKI, the general idea of distributed ledgers, and some specific implementations. Section 3 defines the requirements and choices in an integrated framework used to evaluate and categorize proposed designs. We shortly present the reviewed papers in Sect. 4 and apply the previously defined framework to them. Based on this, in Sect. 5 we state unresolved challenges and provide some recommendations for promising solutions. To sum up, Sect. 6 concludes our work.

2 Background

This section introduces the technical foundations of the Internet X.509 Public Key Infrastructure, and Distributed Ledger Technology (DLT), focusing on Ethereum and smart contracts.

2.1 *Public Key Infrastructure and Internet X.509 Standard*

A Public Key Infrastructure describes a system that can issue, revoke, distribute, and validate certificates. Thereby, a PKI enables entities to securely communicate on an open public network and reliably verify the identity of an entity via digital signatures [1]. The term PKI is used ambiguously, both for the abstract concept as well as for the implementations. As of now, the widely accepted and used implementation of a PKI is the Internet X.509 Public Key Infrastructure, the *current PKI*. In this architecture, clients own a Certificate Trust Store. A Certificate Trust Store manages a list of known digital certificates. For standard end users the Certificate Trust Store contains certificates of the root CAs. These root certificates are used to sign intermediate CAs. CAs verify the connection between an identity and a public key and issue a certificate which can be used to prove this link to others. Clients can also customize their trust by adding specific self-signed certificates. One of the main applications of the PKI is the validation of certificates of domains. During the TLS handshake, domain owners provide the certificate to the client. The client can verify the certificate through trust to any root certificate which is included in its Certificate Trust Store [23].

If the associated key of a certificate is no longer secure, the certificate must be revoked even before the actual expiration date so that the key can no longer be used. Another reason for revocation is incorrect certificate content, for example, in the case of a name change. The current PKI provides multiple techniques for revoking certificates:

2.1.1 Certificate Revocation List

A certificate revocation list (CRL) is a list containing invalid certificates. If a certificate authority (CA) declares a certificate invalid, it enters the serial number of this certificate into the CRL. CRLs are, for example, frequently requested by browsers. The CRLs are parsed whenever the browser validates a particular certificate with a CA, which should be done before any use of the key [16].

One challenge with revocation Lists derives from them being negative and thus can tell one exclusively about the invalidity, rather than the validity of a certificate. In addition, a time gap exists which could allow a revoked certificate to be accepted, since CRLs are cached to avoid the overhead of repeatedly downloading them.

2.1.2 Online Certificate Status Protocol

A more recent protocol is the Online Certificate Status Protocol (OCSP), which is a network protocol that allows clients to query the status of X.509 certificates at a validation service directly at a CA [32].

2.1.3 Certificate Transparency

Certificate Transparency (CT) was introduced by Google to create public certificate logs (CLs) [26]. CAs submit certificates at multiple CLs before they are issued. The CLs generate a Signed Certificate Timestamp (SCT) to prove the consistency of the certificate at a certain timestamp. The CA adds this SCT to the certificate before issuing it to the domain owner. Clients receive the SCT as an extension of the certificate or during the TLS handshake with OCSP. If clients are not able to verify the SCT with a CL they reject the connection to a service. The domain owner should monitor the CL continuously and report maliciously issued certificates. This architecture eases the detection of maliciously issued certificates and revoke them, but due to the rapidly increasing volume of certificates in public logs it is impractical for the domain owner to monitor themselves, and thereby it does not fully prevent man-in-the-middle attacks [27]. Although the use of OSCP and CT provides up-to-date information through direct queries, it adds additional latency during the TLS handshake and can be questionable from a data protection perspective, as CAs or operators of CLs can track which pages are visited.

2.2 Distributed Ledger Technology

Distributed Ledger Technology (DLT) provides a promising platform for addressing the mentioned revocation, security, and privacy issues. Besides, it can optimize certificate revocation and the trust store management of clients.

DLT describes a newly emerged technique in which a set of records, a ledger, is held by multiple entities in a peer-to-peer network. The main advantage of DLT is that no central instance needs to be trusted [17].

A blockchain is a cryptographic implementation of a DLT. The characteristics of distributed and decentralized are enabled by an open system architecture, managed through a consensus mechanism. Making use of append-only blocks as a data structure with Merkle Root trees, records are immutable and easy to verify.

Consensus mechanisms are used for achieving consensus in multi-agent and distributed systems on a single data value or a single state of the network. This plays an important role in non-centralized systems like blockchains. Thereby the consensus mechanism decides which block is added next on the chain. Besides the two famous consensus mechanisms, Proof-of-Work and Proof-of-Stake, many more like Practical Byzantine Fault Tolerance (PBFT), Delegated Proof-of-Stake (DPoS), Ripple, or Tendermint exist. Each has its own characteristics in terms of identity management, energy consumption, and relative ownership to be compromised [38].

Append-only blocks are enabled through linked lists. Thereby, the hash of the previous block is part of the next block. Inside the blocks, data is contained in form of hashes, which are stored in a Merkle hash tree. Those trees can be used to verify if any kind of data presented by a third party is stored on the chain [41].

In the peer-to-peer network, nodes can take on various roles. There are *Full Nodes* that hold the full blockchain, *Mining Nodes* that verify transactions and add them to the chain, and *Wallets* that enable clients with resource-constrained devices to participate in the blockchain network [6].

2.2.1 Ethereum

Ethereum is a DLT that extends the functionality of a blockchain with smart contracts. Smart contracts can execute business logic in the Ethereum Virtual Machine (EVM) which runs on the full nodes of the peer-to-peer network. Due to the increased complexity, the data structure design of the Ethereum blockchain distinguishes between changeable data, which can be changed at any time, and permanent data, which is static as soon as a block is created [41].

Ethereum is currently using the energy-intensive Proof-of-Work consensus mechanism, but it is expected to transition to a Proof-of-Stake consensus mechanism in 2021 [10]. Key to the Proof-of-Work mechanism is an algorithmic puzzle. The answer of this puzzle is easily verifiable, however, challenging to solve. The miner with the correct solution is allowed to add the next block to the blockchain. The severity of these problems in blockchain networks depend on the number of miners and is represented by the difficulty target. This value is set accordingly. The time it takes to mine a new block is called block time [20].

The security of Ethereum has been explored by many pervious works [12, 28]. The threats proposed in these papers are valid, but currently technically difficult to accomplish and usually not economically worthwhile. The developers, a large and open community, discuss and react to addressed issues and thereby ensure the stability and security of the system [18].

2.2.2 Smart Contracts

A smart contract is a piece of code capable of executing predefined instructions. These instructions can be seen as contract content. This could be a verification, an enforcement, a negotiation or an execution of a contract without relying on third parties. There are two different forms of interacting with a smart contract: Retrieving or changing a state of a smart contract. By changing a state a new transaction is created.

The deployed smart contracts and the transactions are irreversible and trackable since they are stored on the Ethereum blockchain. Smart contracts aim to provide higher security compared to the traditional contract law while reducing costs and allowing access for everyone [33].

Smart contracts for the Ethereum blockchain are mostly written in the programming language Solidity, but other languages like Vyper or Bamboo coexist. The smart contract code is compiled to low-level machine instructions (called opcodes) and encoded in bytecode.

The bytecode is sent as a transaction to the Ethereum blockchain. Any changes in the contract code would result in a new byte code that needs to be deployed again and receives a new address. The code is publicly available but the users of the smart contract can be restricted by conditions in the smart contract. Deployed bytecode can be reverse engineered.

Executing functions in smart contracts requires a transaction fee. In the Ethereum blockchain, the fee is called "Gas." This fee differs based on storage, computational complexity, and bandwidth use.

3 Evaluation Framework

This section presents the methodology used for comparing and evaluating the existing blockchain-based PKIs. It consists of two separate parts: the requirements that abstract the whole concept and the architecture decisions that are part of the design stage.

3.1 Requirements

We present ten abstract requirements that a PKI must fulfill to be successful. The requirements are both functional, i. e., what the system should be able to do, and non-functional, i.e., how the system performs these functions. As these requirements are mostly independent from the used technology and implementations, we derive many of our requirements from the NIST Public Key Infrastructure Study published in 1994 [7].

When these original requirements were established, there was no pre-existing global trust infrastructure to replace—a circumstance that has changed since. The available PKIs back then were also designed for more limited resources and connectivity. To account for the changed circumstances and evaluation object, we redefine the requirements with our own description and add three novel requirements which have proven to be important in the past decades. We also exclude some requirements focused on legal matters which were rarely covered in previous work and can therefore not be adequately evaluated.

While defining the requirements, we pay special attention to one use-case: the certification of domain names and the establishment of a TLS connection where only the server is authenticated.

These requirements apply to any blockchain-based PKI. For this evaluation, we assume that any PKI design is capable of certifying identities, distributing the certificate and validating it. Furthermore, the certificates are presumably unforgeable and the system fulfills basic security requirements. Beyond this baseline of functionality, the system should fulfill as many of the following requirements as possible:

Ease of Use: Users only participate in a security system if the effort to do so is low enough. Therefore, obtaining and validating a certificate should be a simple process, fast and not unreasonably expensive. This can be achieved through a simple underlying system or good intermediaries abstracting away some of the complexity as well as a flexible pricing system with low fixed costs.

User Authentication: Before an identity is certified, the user requesting the certificate must be authenticated to prove possession of the identity. This prevents false linkage between certificates and identities. In the context of domain name certification, a proof-of-ownership process for the domain must be executed before the certificate is issued. If this is not the case or the process has weaknesses, unauthorized certificates can be obtained and man-in-the-middle attacks performed.

Certificate Policy: It should be possible to associate a certificate with policies defining the use-cases and domains for which the certificate is applicable, as well as the security requirements the certificate must fulfill [14]. For example, in X.509 certificates the field "KeyUsage" can name nine predefined areas the certificate can be used and accepted for [16].

Trusted Anchors: For any PKI to be effective, the users must be able to trust in some part of it. This part is usually referred to as the anchors. In a hierarchical PKI, the root CAs take on this role, in a web of trust it is decentralized to any participant who certifies its trust in another participant. This trust can then be transferred to other participants to allow the establishment of secure connections. In a blockchain-based PKI, the trust would have to be placed in the distributed ledger itself and by extension the participants running it.

Privacy of Users: Participants might want to make use of a PKIs service without revealing their identity or information about themselves. This requirement is two-fold: The privacy of a user's identity is protected if it is possible to obtain a certificate for the identity without revealing it. The importance of this requirement was proven by attacks enabled by the introduction of certificate transparency: Malicious actors found supposedly private subdomains for the setup of websites in the logs, giving them full control over the website [8]. The second part of this requirement concerns the privacy during certificate validation. A user should not have to reveal the validated identity, as this would allow for the tracking of activity, e. g. visiting a certain domain. The Online Certificate Status Protocol (OCSP) in the current global PKI violates this requirement, as the identity is revealed to the CA that is asked for the certificate status and potential eavesdroppers on the connection.

Certificate Revocation: If a certificate turns out to be compromised or unauthorized, there must be a way to revoke the trust in this certificate. This revocation should become effective as fast as possible to reduce the time frame for potential attacks and it should not require excessive resources. Ideally, the responsible parties should also be able to enforce the delivery of this revocation and reactivate the certificate [34, Ch. 1]. In the current global PKI, this requirement is covered through Certificate Revocation Lists and the OCSP standard.

Liability: In case of a failure of the PKI system, parties which incurred losses due to the failure will want to recuperate their losses. The party responsible and therefore liable for a certain step of the process can be defined in various places in a blockchain-based model: technologically, contractually, or by law. In the current PKI, liability seems to be mostly defined by a contract. However, the wording of the contracts has been criticized as heavily in favor of the CA [31].

Transparency: Since the introduction of certificate transparency logs [26], transparency itself has become something that is expected of a PKI system. It means that a public log of all issued certificates exists, allowing for the faster detection of unauthorized certificates and an audit of the certificates issued by a certain participant.

Scalability: Any proposed implementation or concept must be able to scale up to the standard of the current of the current PKI. It must be able to handle the certification, revocation, and frequent validation of millions of certificates. We review this requirement in three interconnected categories: performance, storage, and cost. Performance mostly concerns the computational power and the time it takes to complete certain actions of the PKI. Storage considers the absolute space occupied by the system as well as the space required on each system. Cost is a combination of the two previous categories as well as other factors which might influence the price attached to using certain functions of the PKI.

Change Incentive: As a global PKI is already established and an integral part of many applications, so for a new system to be adopted it must provide an incentive. This incentive could be a far better fulfillment of the other requirements, e. g. leading to increased security. However, it is conceivable that this advantage is not clear enough or outweighed by the short-term costs of switching between systems. In this case, the system might need to include features designed to convince participants of the current PKI to switch over, e. g. financial incentives. It should be clear though that fulfillment of this requirement first and foremost relies on an overall improved system compared to the current PKI.

3.2 Architecture Decisions

When reviewing the architecture of blockchain-based PKI, we frequently observed certain decisions with regards to scope, trust management, and distributed ledger technology. Based on this observation, we identified three decisions a majority of the reviewed publications made in their architecture:

Scope: The previous works vary widely in their scope and the set of features they include. This has an impact on the fulfillment of functional requirements as well as the ease or difficulty of transitioning to the proposed solution. While each paper has its individual design, we broadly categorize them as follows:

- *Alternative*: The system is designed to replace the current PKI. This usually includes a drastic change to the trust structure or the trust anchors, for example, by moving to a web of trust. Another common design falling into this category are those which combine DNS and PKI into a holistic system, storing both domain ownership and associated certificates on a blockchain.

- *Extension*: The system extends the functionality of the current PKI, for example, by including additional steps in the issuing or validating of certificates. Designs falling into this category are commonly concerned with improving transparency or the efficiency of revocation. They can therefore be compared to the OCSP protocol or certificate transparency logs, which were also added later to the current PKI.

Trust Structure: This concerns how trust is established between the different participants of a PKI. The problem can be depicted as a graph, where the participants are the nodes and the edges are the trust between any two participants. Based on this, two main structures have emerged:

- *Hierarchical Trust Structure*: The graph has one or multiple trust anchors/roots, from which the trust in all other participants is derived. As long as a user trusts these roots, one can trust all participants. This is how the current global PKI is designed, with the trust anchors being the root CAs in the trust store of operating systems and software.
- *Web of Trust*: Any participant can record its trust in another participant. Trust with an unknown participant is established by finding a trust path between the two nodes in the graph. This can be further customized by defining the maximum length of such a trust path.

Blockchain Technology: Any blockchain-based PKI concept must be implemented through an underlying blockchain. The possibilities and features of such a concept are therefore also limited and impacted by this blockchain and it needs to be chosen carefully. Generally speaking, there are two available paths:

- *Existing Blockchain*: The PKI concept can be built on top of an existing distributed ledger, like Bitcoin or Ethereum. This greatly reduces the implementation effort, the blockchain is already proven to work, and it has an existing pool of miners and users. This existing computational power is especially important to prevent threats like the 51% attack [12]. However, using an existing blockchain also means the implementation is limited by the available functionality.
- *Custom Blockchain*: As an alternative, the blockchain could be designed and optimized to support the specific functionality needed for the PKI concept. This would allow pricing, storage, and participation to be tailored for the requirements of the concept.

4 Evaluation

In this section, we shortly present and compare the reviewed papers. In addition, we provide some observations about general trends in the comparison.

4.1 Literature Review

We shortly introduce the reviewed papers, reiterating their focus and ideas. Many of them were identified through a previous comparison by Brunner et al. [9]. We also exclude papers that described concepts not fit for a global PKI, for example, one addressing localized authentication between IoT devices [22].

Fromknecht et al. [19]: The authors focus on the problem of identity retention in PKIs, meaning that only one public key can be associated with an identity at a time. They build their solution on top of the DNS of Namecoin, where only the owner of a domain can publish information for it. The paper discusses recovery after a private key leakage through a second offline key as well as variants of the system which improve efficiency [19].

Wilson and Ateniese [40]: The paper presents several enhancements to PGP by making use of a novel Bitcoin-based PGP certificate format and a design of a distributed PGP key server. They propose to make use of Bitcoin transactions as a distributed PGP key server, using them for certificate storage, retrieval, verification, and revocation [40].

Ali et al. [5]: This paper reports on the experiences made while running Blockstack, a combined DNS and PKI, on the Namecoin and Bitcoin blockchains. The design is extensive, as it links ownership of a domain and publishing certificates for this domain in a single system; the design is also abstracted from the underlying blockchain through a virtual layer. They also mention practical threats posed to the security of the system due to the properties of the underlying blockchain [5].

Matsumoto and Reischuk [30]: This paper proposes a design called Instant Karma PKI (IKP). It is a blockchain-based PKI enhancement that puts a focus on automatic responses to CA misbehavior and incentives for those who help detect misbehavior. They extensively analyze their approach with methods of game theory [30].

Al-Bassam [4]: A decentralized public key infrastructure system that utilizes the transparency of the blockchain was developed in this work. It is built around fine-grained attribute management making use of a web of trust approach. The costs associated with using smart contracts are evaluated for certain amounts of storage and functionalities. It points out that the adoption barrier of a web of trust system is high [4].

Wan et al. [36]: This paper combines the proposal of requiring certificates to be signed by multiple certificate authorities with a blockchain-based log of domain names and certificates. The presented design is abstract and agnostic of specific blockchain implementations, but a prototype is created on Ethereum to evaluate the performance [36].

Ahmed and Aura [3]: The authors focus on a formal definition of the policies that domain owners, trust store owners, and CAs can register on a blockchain and certificates can reference. They seek to make trust structures more flexible by allowing the previous parties to revoke trust in each other at any time. They make extensive use of the subscriber-observer pattern in Ethereum smart contracts to implement these connections [3].

Chen et al. [11]: This paper proposes a blockchain-based audit scheme for certificates called CertChain. They make use of a custom blockchain. By introducing dual counting Bloom filters, the approach enables a low-memory architecture and efficient queries to check the validity of a certificate [11].

Yakubov et al. [42]: The design seeks to combine certificate transparency in smart contracts on a blockchain. It proposes that all CAs have smart contracts which hold the certificates and their status. The location of this information would be embedded in X.509 certificate extension fields, allowing a recipient to verify it [42].

Khieu and Moh [24]: This design, called Cloud-based PKI, is similar to Yakubov (2018), also covering certificate transparency and revocation through smart contracts. In a similar fashion to current certificate transparency, certificate verification would include retrieving its status from the blockchain to see if it is valid and not revoked. Furthermore, the proposal leverages cloud providers to make the blockchain more resilient to attacks on its availability [24].

Kubilay et al. [25]: Their solution named CertLedger is a holistic design, introducing blockchain technology to both certificate issuance and revocation. They provide a complete specification of the design and evaluate it. A start-up[1] based on the described system was launched by the authors [25].

Chiu et al. [13]: This paper explains the design of a decentralized PKI system, called NoPKI, which has multiple small, dynamically formed, and trusted groups called the neighborhood. Thereby each interaction (registration, revocation, and validation) with the PKI requires witnesses of neighbors. They implemented their design and evaluated long-term accessibility, certificate security, and public auditability purposes [13].

Wang et al. [39]: This approach seeks to balance the power of CAs by requiring certificates and revocation to be recorded in a public blockchain. A user of the blockchain can take ownership of a domain name through approval by other domain owners, and then perform certificate actions for this domain. This idea could be an addition to the already existing PKI [39].

Toorani and Gehrmann [35]: This detailed, but abstract design uses a decentralized approach on a custom blockchain to enable certificate validation. A certificate can only be registered or revoked through the consensus of the group. They pay special attention to the scalability of certificate verification through an accumulator, but do not perform an experimental evaluation [35].

[1] https://certledger.io/.

4.2 Comparison

Now we apply the framework to the reviewed designs. The result is presented in Table 1. When considering the fulfillment of requirements, they can be completely fulfilled (✓), partially fulfilled (□), e. g. if a requirement has multiple aspects and only some are met, or not fulfilled (✗). If the implementation described in the paper does not cover the area relevant for evaluating a requirement, it is marked as not answered (-).

For non-functional requirements (Ease of Use, Trusted, Privacy, Scalability), the fulfillment is considered relative to the current global PKI. If the fulfillment is comparable to the current PKI or better, the requirement is considered as met.

Since the reviewed literature was almost purely focused on the technological challenges, we were not able to evaluate the requirement *Liability*. The change incentive requirement faces a similar problem. In addition to this, to evaluate the change incentive, the fulfillment of the requirement needs to take the whole system into account. However, the different systems vary so much in scope that they cannot be adequately compared. Hence, we decided not to evaluate the *Change Incentive* as well.

The last requirement, scalability, is evaluated for each of the three aspects: performance, mostly regarding the speed of executing operations, storage, and cost. Few designs evaluated all these aspects and even fewer considered them with a realistic number of certificates and domains. Therefore, we had to extrapolate the numbers and can only judge them by orders of magnitude.

4.3 Discussion

We would first like to clarify that a low fulfillment of requirements does not indicate the proposed designs are flawed. Many of the existing solutions were not specifically designed to handle large-scale domain name validation. We still decided to include them as they might offer innovative solutions to the open challenges posed by the different requirements. This variation in scope is also a major reason as to why so many requirements are indicated as not answered (-) in the comparison table.

During the evaluation of the designs, some incompatibilities between the framework and the papers became noticeable. Namely, the requirement *Ease of Use* was tough to evaluate as systems usually become more accessible by creating an environment of supporting services. For example, these could be intermediary services which remove some of the complexity of obtaining a certificate. However, this environment usually develops naturally due to an unfilled demand, rather than due to planning by the creators of the underlying system. We can therefore only evaluate if the proposed design would inhibit the establishment of such intermediaries in any way.

Table 1 Comparison of blockchain-based PKIs using the framework from Sect. 3

Paper	Scope	Trust structure	Blockchain	Ease of use	User authentication	Certificate policy	Trusted anchors	Privacy validation/ certification	Revocation	Transparency	Scalability performance/ storage/cost
Fromknecht et al. [19]	Alternative	–[a]	Namecoin	-	✓	✗	✗	✓ / ✗	✓	✓	✓ / ✓ / -
Wilson and Ateniese [40]	Alternative	Web of Trust	Bitcoin	✗	✓	-	✓	- / ✗	✓	✓	✗ / - / -
Ali et al. [5]	Alternative	–[a]	Bitcoin	✓	✓	-	✗	- / ✗	✓	✓	✗ / □ / -
Matsumoto and Reischuk [30]	Extension	Hierarchical	Ethereum	✓	✓	✓	✓	✓ / ✗	-	✗	- / - / ✓
Al-Bassam [4]	Alternative	Hierarchical	Ethereum	✗	✓	-	✓	- / ✗	✓	✓	✗ / □ / ✓
Wan et al. [36]	Extension	Hierarchical	Ethereum	✓	✓	-	✓	✗ / ✗	✓	✓	- / - / -
Ahmed and Aura [3]	–[b]	Hierarchical	Ethereum	-	✓	✓	-	- / ✗	-	✗	✗ / ✓ / ✓
Chen et al. [11]	Extension	Hierarchical	Custom	✓	✓	-	✓	✗ / ✓	✓	✗	✓ / ✓ / ✗
Yakubov et al. [42]	Extension	Hierarchical	Ethereum	✓	-	-	✓	- / ✓	✓	✗	- / ✓ / ✓
Khieu and Moh [24]	Extension	Hierarchical	Ethereum	✓	-	-	✓	- / ✓	✓	✗	✓ / - / □
Kubilay et al. [25]	Alternative	Hierarchical	Ethereum	✓	✓	✓	✗	✓ / ✗	✓	✓	✓ / ✗ / □
Chiu et al. [13]	Alternative	Web of Trust	Custom	✓	✓	✗	✗	- / ✗	✓	-	✓ / - / -
Wang et al. [39]	Extension	Hybrid	Custom	✗	□	-	✓	✓ / ✗	✓	✓	✓ / ✓ / -
Toorani and Gehrmann [35]	Alternative	Web of Trust	Custom	-	✓	-	✓	- / ✓	✓	✗	✓ / - / -

[a] Domain registration is part of the system, so proof-of-ownership is provided by the blockchain rather than a trust anchor
[b] Abstract concept that needs to be part of another system

Another issue is the mutual exclusivity of the requirements *Transparency* and *Privacy of Certification*: while it is possible to fulfill neither, it is often impossible to fulfill both at the same time. Because of this exclusivity it becomes impossible to meet all requirements.

The definition for evaluating the requirement *Scalability* can also be improved further. So far, the evaluation was mostly comparative and based on the subjective judgement of the reviewed designs. A common standard for the time performance and costs of certain functions as well as the overall storage space should be defined. This standard could then be applied evenly to the numbers reported by the different papers. However, it would have to be flexible to account for the different circumstances of the scenarios under which the papers evaluated their scalability.

On a high level, some trends are noticeable in the comparison table. In the following, we seek to provide some insights into the causes for these trends, both based on the reasoning of the reviewed designs and our own observations. Trends, which directly lead to a recommendation, will be discussed in the next section.

With regards to the architecture decisions, a clear trend of making the same decisions as the current PKI can be found. The current global PKI is hierarchical and transparent, which a majority of the proposed solutions shares. Similarly, the selected blockchain technology is often an existing one; this reduces the work for specification and implementation and allows for an easier evaluation under real-world circumstances. Ethereum is by far the most popular choice, likely due to the availability of smart contracts in combination with its size and popularity.

About half of the proposed solutions also do not seek to replace the current PKI; rather, they want to add onto it and improve it similarly to certificate transparency. This can be an additional verification of certificates before they are published as proposed in Wang et al. [39] or a dynamic audit of certificates before they are accepted for a TLS connection [30].

Regarding the requirements, most implementations seem to be focused on functionality. Essentially all provide user authentication and certificate revocation and most provide some form of transparency and therefore the possibility of audits. The designs only start to show their differences in the non-functional requirements.

5 Open Issues

In this section, we discuss the problem of scalability, holistic solutions and the transition from the current PKI to a blockchain-based PKI.

5.1 Scalable Design

While the reviewed designs bring many innovative solutions to the table to solve the problem of scalability, none can with certainty fulfill all three aspects—

performance, storage, cost. This is also an inherent problem of blockchain-based technology: due to the decentralization creating many replicas of the data, storing information in a simple database is always cheaper—at much lower integrity guarantees though.

However, if no solution can fulfill all scalability requirements at once yet, the solution may lay in merging them. Some of the most promising ideas are the creation of separate chains for different validity periods, the secure discarding of some content of older blocks [39], or the creation of a virtual chain which only uses the underlying blockchain for integrity, but not storage [5]. An approach combining the best ideas from these papers should be pursued and evaluated.

This evaluation should be performed with a realistic number of certificates and revocations, to ensure the system is up to the current standard. Furthermore, one of the reasons cited for having to replace the current PKI is its inability to keep up with the demand of future technologies like IoT [24]. This should also be considered in any evaluation to ensure the system is future-proof.

Finally, while a custom blockchain as the base technology for a system has such advantages as the ability to fine-tune its features for the requirements, it also makes it harder to evaluate the cost aspect. The four reviewed papers with custom blockchains did not evaluate the cost scalability [13, 35, 37, 39]. One reason for this could be the lack of an exchange rate between the new blockchain and any other crypto or fiat currency. Nevertheless, it should be possible to transfer the knowledge about operating costs on existing blockchains to new ones and provide an assumption for the cost.

5.2 Holistic Solutions

Merging the ideas of multiple designs would also address another issue: most of the proposed solutions only address a part of the requirements and issues. This is clearly visible in Table 1 with the many requirements that are marked as not answered as the paper does not concern itself with this specific aspect. While this is understandable due to the limited scope of research projects, for any system to be adopted on a larger scale it needs to be specified in full.

Especially for the designs wanting to fully replace the current PKI, it is therefore crucial to present a holistic solution. An evaluation of the feasibility and impacts of such an alternative is also only possible with a complete design. Otherwise, some assumptions for missing aspects always need to be made that might not hold up in reality.

Another approach might be to improve the modularity of the different functionalities a PKI offers. As of now, the functions of issuance, validation, revocation, and transparency are already split across multiple protocols and parties. A definition of the different functions and their interfaces would allow for partial solutions to be designed and implemented.

No matter the approach, any proposed solution should clearly state the functions it wants to cover and how the interaction with other parts of the PKI system –if they presumably exist– would be handled.

5.3 *Transition*

While technical challenges are the focus of many papers on this subject, the transition from the current PKI to a distributed ledger-based PKI in political and economic dimensions is rarely addressed in the research landscape. Once there are holistic solutions, questions about political and economic change management become urgent.

From a political and organizational point of view, standardization and the adoption process need to be worked out and agreed upon. Organizations must be appointed to steer this process and mediate between the parties involved to establish a standard. Core issues that cannot be addressed technically, such as liability over errors in new systems, must be discussed. The process must be broken down into milestones and the speed of implementation must be determined. Besides, it must be discussed whether and how to ensure compatibility with old systems.

From an economic point of view, financial change incentives must be specified for entities of the current PKI. For this purpose, user stories have to be written from the perspective of the end users, outlining their financial interests. This must also include the environmental costs. Additionally, the system must be designed so that it supports correct behavior and penalizes misbehavior.

6 Conclusion

In this paper, we proposed a comprehensive evaluation framework for public key infrastructures on a distributed ledger, including a detailed description of the requirements. We reviewed fourteen existing solutions and applied the evaluation framework to them. Also, we discussed the general trends in these solutions and the application of the framework. Furthermore, we proposed an enhanced hybrid trust infrastructure and highlighted work that needs to be addressed in the near future. We suggested that combining various approaches solves final technical problems in the area of scalability, and research on a smooth transition is necessary for adoption.

The proposed framework can be used to identify weak points in existing and future blockchain-based PKI designs and to ensure their feasibility. To further improve the framework, the requirements could be defined quantitatively rather than qualitatively, leading to more objective results.

References

1. Adams, C., Lloyd, S.: Understanding PKI: Concepts, Standards, and Deployment Considerations, 2nd edn. Addison-Wesley Longman Publishing Co. Inc., Boston (2002)
2. Adkins, H.: An update on attempted man-in-the-middle attacks. https://security.googleblog.com/2011/08/update-on-attempted-man-in-middle.html. Accessed 09 Mar 2021
3. Ahmed, A.S., Aura, T.: Turning trust around: smart contract-assisted public key infrastructure. In: 2018 17th IEEE International Conference On Trust, Security and Privacy In Computing And Communications/12th IEEE International Conference On Big Data Science And Engineering (TrustCom/BigDataSE), pp. 104–111. IEEE, Piscataway (2018)
4. Al-Bassam, M.: SCPKI: a smart contract-based PKI and identity system. In: Proceedings of the ACM Workshop on Blockchain, Cryptocurrencies and Contracts, pp. 35–40 (2017)
5. Ali, M., Nelson, J., Shea, R., Freedman, M.J.: Blockstack: a global naming and storage system secured by blockchains. In: 2016 Annual Technical Conference 16, pp. 181–194 (2016)
6. Antonopoulos, A.M.: Mastering Bitcoin: Unlocking Digital Crypto-Currencies, 1st edn. O'Reilly Media Inc., Sebastopol (2014)
7. Berkovits, S., Chokhani, S., Furlong, J.A., Geiter, J.A., Guild, J.C.: Public key infrastructure study. Technical report. National Inst of Standards and Technology, Gaithersburg (1994)
8. Böck, H.: Hacking web applications before they are installed (2017). https://www.golem.de/news/certificate-transparency-hacking-web-applications-before-they-are-installed-1707-129172.html. Accessed 11 Mar 2021
9. Brunner, C., Knirsch, F., Unterweger, A., Engel, D.: A comparison of blockchain-based pki implementations. In: Proceedings of the 6th International Conference on Information Systems Security and Privacy (ICISSP 2020), pp. 333–340 (2020)
10. Buterin, V., Conner, E., Dudley, R., Slipper, M., Norden, I., Bakhta, A.: EIPs/eip-1559.md at master · ethereum/EIPs · GitHub. https://github.com/ethereum/EIPs/blob/master/EIPS/eip-1559.md. Accessed 14 Mar 2021
11. Chen, J., Yao, S., Yuan, Q., He, K., Ji, S., Du, R.: CertChain: public and efficient certificate audit based on blockchain for TLS connections. In: IEEE INFOCOM 2018-IEEE Conference on Computer Communications, pp. 2060–2068. IEEE, Piscataway (2018)
12. Chen, H., Pendleton, M., Njilla, L., Xu, S.: A survey on ethereum systems security: vulnerabilities, attacks, and defenses. ACM Comput. Surv. **53**(3) (2020). https://doi.org/10.1145/3391195
13. Chiu, W.Y., Meng, W., Jensen, C.D.: NoPKI – a point-to-point trusted third party service based on blockchain consensus algorithm. In: Communications in Computer and Information Science, pp. 197–214. Springer, Singapore (2020). https://doi.org/10.1007/978-981-15-9739-8%5F16
14. Chokhani, S., Ford, W., Sabett, R., Merrill, C., Housley, R., Wu, S.: Internet X.509 Public Key Infrastructure Certificate Policy and Certification Practices Framework. RFC 3647, RFC Editor (2003). https://tools.ietf.org/html/rfc3647
15. Comodo: Report of incident on 15-MAR-2011. https://www.comodo.com/Comodo-Fraud-Incident-2011-03-23.html. Accessed 09 Mar 2021
16. Cooper, D., Santesson, S., Farrell, S., Boeyen, S., Housley, R., Polk, W.: Internet X.509 Public Key Infrastructure Certificate and Certificate Revocation List (CRL) Profile. RFC 5280, RFC Editor (2008). https://tools.ietf.org/html/rfc5280
17. Deshpande, A., Stewart, K., Lepetit, L., Gunashekar, S.: Understanding the landscape of Distributed Ledger Technologies/Blockchain: Challenges, opportunities, and the prospects for standards. Understanding the landscape of Distributed Ledger Technologies/Blockchain: Challenges, opportunities, and the prospects for standards (2020). https://doi.org/10.7249/rr2223
18. Ethereum Community: Home | Ethereum Improvement Proposals. https://eips.ethereum.org/. Accessed 21 Mar 2021
19. Fromknecht, C., Velicanu, D., Yakoubov, S.: A decentralized public key infrastructure with identity retention. IACR Cryptology ePrint Archive, p. 803 (2014)

20. Gervais, A., Karame, G.O., Wüst, K., Glykantzis, V., Ritzdorf, H., Čapkun, S.: On the security and performance of proof of work blockchains. In: Proceedings of the ACM Conference on Computer and Communications Security, pp. 3–16 (2016). https://doi.org/10.1145/2976749.2978341
21. Goodin, D.: Google takes Symantec to the woodshed for mis-issuing 30,000 HTTPS certs (2017). https://arstechnica.com/information-technology/2017/03/google-takes-symantec-to-the-woodshed-for-mis-issuing-30000-https-certs/. Accessed 09 Mar 2021
22. Hammi, M.T., Hammi, B., Bellot, P., Serhrouchni, A.: Bubbles of trust: a decentralized blockchain-based authentication system for IoT. Comput. Secur. **78**, 126–142 (2018)
23. Housley, R., Ford, W., Polk, T., Solo, D.: Internet x.509 public key infrastructure certificate and CRL profile. RFC 2459, RFC Editor (1999). http://www.rfc-editor.org/rfc/rfc2459.txt
24. Khieu, B., Moh, M.: CBPKI: cloud blockchain-based public key infrastructure. In: Proceedings of the 2019 ACM Southeast Conference, pp. 58–63 (2019)
25. Kubilay, M.Y., Kiraz, M.S., Mantar, H.A.: CertLedger: a new PKI model with Certificate Transparency based on blockchain. Comput. Secur. **85**, 333–352 (2019)
26. Laurie, B., Langley, A., Kasper, E.: Certificate Transparency. RFC 6962, RFC Editor (2013). https://tools.ietf.org/html/rfc6962
27. Li, B., Lin, J., Li, F., Wang, Q., Li, Q., Jing, J., Wang, C.: Certificate transparency in the wild: exploring the reliability of monitors. In: Proceedings of the 2019 ACM SIGSAC Conference on Computer and Communications Security, pp. 2505–2520. CCS '19, Association for Computing Machinery, New York (2019). https://doi.org/10.1145/3319535.3345653
28. Li, X., Jiang, P., Chen, T., Luo, X., Wen, Q.: A survey on the security of blockchain systems. Fut. Gen. Comput. Syst. **107**, 841–853 (2020). https://doi.org/10.1016/j.future.2017.08.020
29. MarketsandMarkets: Public Key Infrastructure Market by Component (Solution and Services (Professional and Managed)), Deployment Type, Organization Size, Vertical (BFSI, Telecom and IT, Healthcare and Life Sciences), and Region – Global Forecast to 2024. https://www.marketsandmarkets.com/Market-Reports/public-key-infrastructure-market-145372975.html. Accessed 09 Mar 2021
30. Matsumoto, S., Reischuk, R.M.: IKP: turning a PKI around with blockchains. IACR Cryptology ePrint Archive **2016**, 1018 (2016)
31. Roosa, S.B., Schultze, S.: The "Certificate Authority" trust model for SSL: a defective foundation for encrypted Web traffic and a legal quagmire. Intellect. Prop. Technol. Law J. **22**(11), 3 (2010)
32. Santesson, S., Myers, M., Ankney, R., Malpani, A., Galperin, S., Adams, C.: X.509 internet public key infrastructure online certificate status protocol – ocsp. RFC 6960, RFC Editor (2013). http://www.rfc-editor.org/rfc/rfc6960.txt
33. Szabo, N.: Formalizing and securing relationships on public networks. First Monday (1997)
34. The Open Group: Open Group Guide - Architecture for Public-Key Infrastructure (APKI). https://pubs.opengroup.org/onlinepubs/9219899/front.htm. Accessed 18 Mar 2021
35. Toorani, M., Gehrmann, C.: A decentralized dynamic pki based on blockchain. In: Proceedings of the 36th Annual ACM Symposium on Applied Computing, pp. 1646–1655 (2021)
36. Wan, Z., Guan, Z., Zhuo, F., Xian, H.: BKI: Towards accountable and decentralized public-key infrastructure with blockchain. In: International Conference on Security and Privacy in Communication Systems, pp. 644–658. Springer, Berlin (2017)
37. Wang, J., Li, S., Wei, S.: Identity-based cross-domain authentication by blockchain via pki environment. In: CCF China Blockchain Conference, pp. 131–144. Springer, Berlin (2019)
38. Wang, Y., Cai, S., Lin, C., Chen, Z., Wang, T., Gao, Z., Zhou, C.: Study of blockchains's consensus mechanism based on credit. IEEE Access **7**, 10224–10231 (2019). https://doi.org/10.1109/ACCESS.2019.2891065
39. Wang, Z., Lin, J., Cai, Q., Wang, Q., Zha, D., Jing, J.: Blockchain-based certificate transparency and revocation transparency. IEEE Trans. Depend. Sec. Comput. **19**, 681–697 (2020)
40. Wilson, D., Ateniese, G.: From pretty good to great: enhancing pgp using bitcoin and the blockchain. In: International Conference on Network and System Security, pp. 368–375. Springer, Berlin (2015)

41. Wood, G.: Ethereum: a secure decentralised generalised transaction ledger. In: Ethereum Project Yellow Paper, pp. 1–32 (2014)
42. Yakubov, A., Shbair, W., Wallbom, A., Sanda, D., et al.: A blockchain-based pki management framework. In: The First IEEE/IFIP International Workshop on Managing and Managed by Blockchain (Man2Block) colocated with IEEE/IFIP NOMS 2018, Taipei, 23–27 April 2018 (2018)

On the Performance of MIMO Vehicular Visible Light Communications

Selma Yahia, Yassine Meraihi, Asma Benmessaoud Gabis, and Amar Ramdane-Cherif

1 Introduction

Over the past couple of decades, vehicle density and volume have increased dramatically, leading to a significant increase in congestion and accidents, with serious impacts on the environment, the economy, and human's quality of life [1]. Intelligent Transportation Systems (ITS) have received much attention as an effective way to improve road safety, travel efficiency, and passenger comfort [2]. As the cornerstone for ITS, vehicular communication includes a number of moving vehicles that establish wireless communications with each other and with the infrastructures along the road. The resultant connectivity forms are vehicle-to-vehicle (V2V), vehicle-to-infrastructure (V2I), and infrastructure-to-vehicle (I2V), which can be commonly referred to V2X communications [3]. Current deployments for V2X applications mainly depend on the use of radio-based solutions such as dedicated-short-range-communication (DSRC) [4] and cellular-V2X communication [5]. In the near future, the widespread prevalence of ITs is highly expected, causing substantial interference levels in the corresponding limited radio frequencies. This often results in a low packet reception rate, especially in heavy traffic [6]. Therefore, an alternative or complementary solution to radio frequency (RF) technologies becomes necessary.

S. Yahia (✉) · Y. Meraihi
LIST Laboratory, University of M'Hamed Bougara Boumerdes, Boumerdes, Algeria
e-mail: s.yahia@univ-boumerdes.dz

A. B. Gabis
Ecole nationale Supérieure d'Informatique, Laboratoire des Méthodes de Conception des Systèmes, Oued-Smar, Alger, Algeria

A. Ramdane-Cherif
LISV Laboratory, University of Versailles St-Quentin-en-Yvelines, Velizy, France

M. D. Hina et al. (eds.), *Advances in Computational Intelligence and Communication*, EAI/Springer Innovations in Communication and Computing,
https://doi.org/10.1007/978-3-031-19523-5_4

Visible light communication (VLC) is one of the most interesting solutions [7] which depends on the simultaneous use of light-emitting-diodes (LEDs) as wireless transmitters and lighting devices. The increasing adoption of LEDs in current automotive lighting such as head- and tail- lights (HLs and TLs), turn signals, and brake lights make the usage of VLC technology a viable option for vehicular communication [8, 9]. In comparison to indoor VLC counterparts, vehicular VLC faces additional challenges [10] such as the effect of mobility [11, 12], exposure to sunlight [13], weather conditions [14], and asymmetrical pattern of outdoor lights [15, 16]. Consequently, extensive research efforts have directed to study different aspects in the vehicular VLC such as the development of dedicated channel models [17–19], Physical layer design [20–22], and investigating the medium access control (MAC) algorithms [23–26]. The possibility of multi-direction reception is also investigated utilizing multiple photodetectors (PDs) either in angle or space diversity forms [27, 28].

Considering the V2V work, which is the focus of this paper, all previous studies have considered single transmission schemes. To enhance the data rates, however, the multiple-input-multiple-output (MIMO) scheme should be considered as one of the effective techniques used to achieve high data rates. In vehicular networks, there are usually multiple sources of LEDs, so it is natural to use MIMO technology to have parallel data transmission and improve the data rates. There are already some efforts suggesting MIMO-VLC systems in Vehicular VLC [29–33]. However, some of these works [31, 33] have used the ideal Lambertian pattern to model the car headlights, which does not reflect the reality since the practical headlights have asymmetric radiation patterns. Furthermore, most of these works consider the only case where the cars move perfectly. In other words, no lateral offset between the cars is considered as in [29, 30, 32]. The impact of vertical and lateral shifts same as the effect of adverse weather conditions should be taken into account for precise analysis as in real vehicular scenarios.

In this paper, we investigate the performance of the MIMO V2V VLC system where two vehicles follow each other on a single-lane road. We construct a 2×2 MIMO transmission system utilizing two HLs of the source vehicle as wireless transmitters while two PDs are installed at the destination vehicle acting as wireless receivers. For the channel modeling, we adopt the ray-tracing features of OpticStudio, validated in [34], taking into account the asymmetrical pattern of the vehicle HLs, the possibility of both vertical and horizontal displacements between vehicles same as the effect of weather conditions. The received optical power and the bit-error-rate (BER) are investigated for each MIMO link between transmitters and receivers.

The structure of this paper is as follows: Sect. 2 describes the channel and system model. Section 3 illustrates the performance analysis. Section 4 presents the numerical results and discussions. Finally, concluding remarks are given in Sect. 5.

2 V2V System and Channel Model

2.1 System Model

As shown in Fig. 1, we consider a V2V scenario in a single-lane road with a width of W_l. The two cars follow each other with a separation distance of d_x and a possible lateral offset of d_h between their centers. A MIMO transmission scheme with N_t transmit LEDs and M_r receive PDs is considered. Specifically, the two HLs of the source car are adopted as wireless transmitters which are denoted by TX1 and TX2 and emitting a total optical power of P_t. On the other side, a pair of PD receivers, denoted by RX1 and RX2, are served as wireless receivers which are placed at the rear of the destination car, under the TLs. Both the two TXs or the two RXs are separated horizontally by a distance d_c from each other. Each PD is characterized by a field of view ψ and an aperture diameter D_r.

In our work, we consider a Spatial Multiplexing (SMP) transmission scheme. The idea of this technique is the simultaneous transmission of independent data streams by N_t LEDs. It provides an enhanced spectral efficiency of $N_t \log_2(R)$ bit/s/Hz, where R is the constellation points [35].

2.2 Channel Model

In this study, we followed the methodology applied in [36]. In which, the non-sequential ray-tracing features of OpticStudio® software was adopted for channel modeling. This approach permits the propagation of rays in any order through the environment, which allows the scattering and reflection of these rays towards any object encountered. In the first step, a 3D test environment model is built in the OpticStudio® platform. In the next step, CAD models of cars are imported and the surface coating materials are then specified. The characteristics of the light source, including the number of emission rays, the optical strength, the radiation pattern, and directions are defined. Besides, receiver parameters such as aperture diameter, active region, field of view, and orientations are also included. The weather

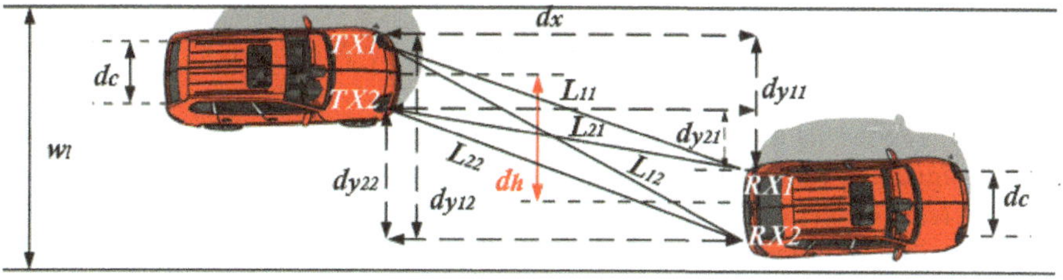

Fig. 1 V2V-VLC system under consideration

condition is also specified using Mie scattering model [36], where the density and radius of particles and the refractive index are provided as input parameters. In the simulated environment, the power, path length, and the propagation delay of each ray emitted between the transmitter and the receiver are computed using the non-sequential ray-tracing method. The rays statistics are then imported into Matlab® software for further processing. Thus, the channel impulse response for each link in the MIMO system can be given by the following equation:

$$h_{ij}(t) = \sum_{k=1}^{K_{ij}} P_{ij}(k)\delta\left(t - \tau_{ij}(k)\right) \tag{1}$$

where K_{ij} is the number of rays emitted from the ith transmitter and captured by the jth receiver. Also, τ_{ij}, P_{ij}, and δ denote the propagation delay, the optical power, and the Dirac delta function, respectively.

Let $L_{ij} = \sqrt{{d_x}^2 + d_{y_{ij}}^2}$ denotes the transmission distance between the ith TX and the jth RX where $d_{y_{ij}}$ is the lateral shift between the ith TX and the jth RX (see Fig. 1). Thus, the channel path loss between the ith TX and the jth RX ($i = 1, 2$ and $j = 1, 2$) can be expressed as [36]

$$H(i, j) = \left(\frac{D_r\left(d_x/L_{ij}\right)^{1/\varepsilon}}{\zeta L_{ij}}\right)^2 \exp\left(-cL_{ij}\left(\frac{D_r}{\zeta L_{ij}}\right)^{\varepsilon/2}\right) \tag{2}$$

where ζ and ε are correction coefficients and c denotes the extinction coefficient for a specific weather type. The associated values for ζ, ε, and c are given in [36].

3 Performance Metrics

In this section, we present the performance analysis of our proposed system. We consider the BER as a performance metric. For SMP, the BER can be expressed as [37]

$$\mathrm{BER_{SMP}} \le \frac{1}{N_t}\sum_{i=1}^{N_t} \frac{2\left(M_i - 1\right)}{M_i \log_2\left(M_i\right)} Q\left(\frac{1}{M_i - 1}\sqrt{\frac{E_s}{N_0 N_t^2 \left\|\mathbf{W}_i\right\|^2}}\right) \tag{3}$$

where E_s represents the mean electrical energy emitted of the intensity modulated optical signals. It can be given by $E_s = (\varrho I)^2 T_s$, here, ϱ symbol denotes the optical to electrical conversion coefficient, I is the average optical power emitted, and T_s is the symbol duration. In (3), N_0 is the spectral density of noise and W_i is the ith row of the Pseudo-inverse of channel matrix H which can be written as $\mathbf{W} = \left(\mathbf{H}^T\mathbf{H}\right)^{-1}\mathbf{H}^T$. Also, M is the selected modulation level.

Table 1 Main simulation parameters for V2V system

System parameters	Values
Road parameters	
Lane width, W_l	3.75 m
Number of lanes	1
Transmitter parameters	
Number of TX, N_t	2
Power, P_t	10 W
Receiver parameters	
Number of RX, M_r	2
Aperture diameter, D_r	1 cm, 2.5 cm, and 4 cm
Lateral shift, d_h	0 m, 1 m, and 2 m
Noise parameters	
Bandwidth, B	20 MHz
Spectral density, N_0	10^{-21} A^2/Hz

4 Simulation Results and Discussion

In this section, we present the numerical results of the V2V system under consideration. As shown in Fig. 1, we consider a V2V scenario in a single-lane road with a width of $W_l = 3.75$ m. We assume that two cars follow each other with a lateral offset of $d_h = 0$ m, 1 m, and 2 m. The total transmit optical power of $P_t = 10$ W is assumed. The aperture diameter of each PD is taken as $D_r = 1$ cm, 2.5 cm, 4 cm. All simulation parameters are included in Table 1. We firstly present the received optical power versus distance for all scenarios under consideration. Then, we present the BER performance versus Receiver SNR (i.e., $SNR_{TX} = E_s/N_0$).

In Fig. 2, we consider the clear weather condition and the receiver aperture of $Dr = 1$ cm to study the effect of misalignment (lateral shift) between cars considering different d_h values. It is observed from Fig. 1a that the received power decreases with increasing inter-vehicular distance. This is due to the attenuation of the light signal as it travels through the air is very large. For example, consider the case of $d_h = 0$ m and the link between TX1-RX1. The received power at $d_x = 10$ m is 6 dBm. This reduces to −2 dBm and −8 dBm for $d_x = 25$ m and $d_x = 50$ m, respectively. It is also obvious that there is a noticeable difference in the receiver power values between different MIMO links for short distances. This is due to the difference in the horizontal locations even the two cars are perfectly aligned. For example, consider $d_h = 0$ m and $d_x = 25$ m. The received power for the TX1-RX1 or TX2-RX2 is −2 dBm while the received power for TX1-RX2 or TX2-RX1 link is given as −3.2 dBm. We can notice that the highest amount of power are received by links TX1-RX1 and TX2-RX2 with a large difference of 8 dBm at $d_x = 10$ m. This difference becomes very small and negligible as the distance increases (i.e., about 0.4 dBm at $d_x = 50$ m). In order to further clarify the effect of lateral shift, we consider the TX1-RX1 link and the distance of $d_x = 10$ m. The received power

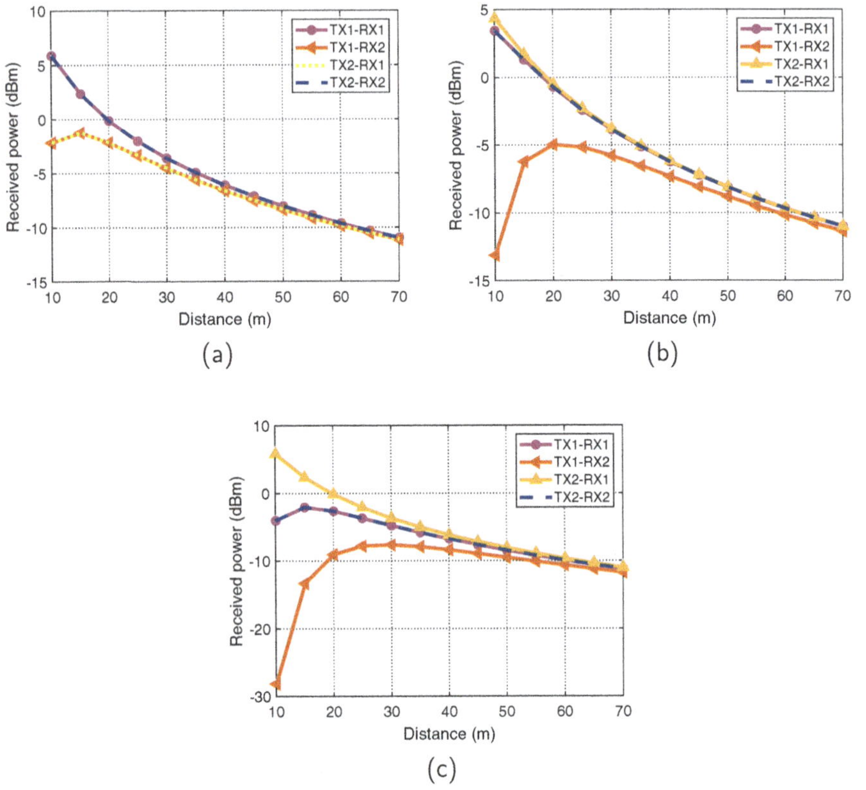

Fig. 2 Received power versus distance for: (**a**) $d_h = 0$ m (**b**) $d_h = 1$ m (**c**) $d_h = 2$ m, for $Dr =$ 1 cm and clear weather

at $d_h = 0$ m is 6 dBm as shown in Fig. 2a. This reduces to 3.5 dBm and -4 dBm for $d_h = 1$ m (see Fig. 2b?) and $d_h = 2$ (see Fig. 2c), respectively.

Figure 3 illustrates the effect of receiver aperture (Dr) on the received power under the assumption of the clear weather condition and the perfect alignment case, i.e., $d_h = 0$ m. It is observed that the received power significantly increases as the aperture diameter increases. This is due to the fact that the amount of the light intensity (number of rays) captured by the receiver increases when the receiving area increases. For instance, consider $d_x = 20$ m, $d_h = 0$ m and the link between TX1-RX1. The received power at D_r 1 cm is 0 dBm as shown in Fig. 3a. This climbs to 8 dBm (see Fig. 3b) and 12 dBm (see Fig. 3c) for $D_r = 2.5$ cm and $D_r = 4$ cm, respectively.

In Fig. 4, we present the corresponding received power of the MIMO links under different weather conditions assuming $Dr = 1$ cm and $d_h = 0$ m. It is observed a small effect for shorter distances but it is significant when the distance becomes sufficiently large. For example, consider $d_x = 10$ m, the received power for clear weather is 6 dBm. This reduces to 5.6 dBm and 5.5 dBm for moderate

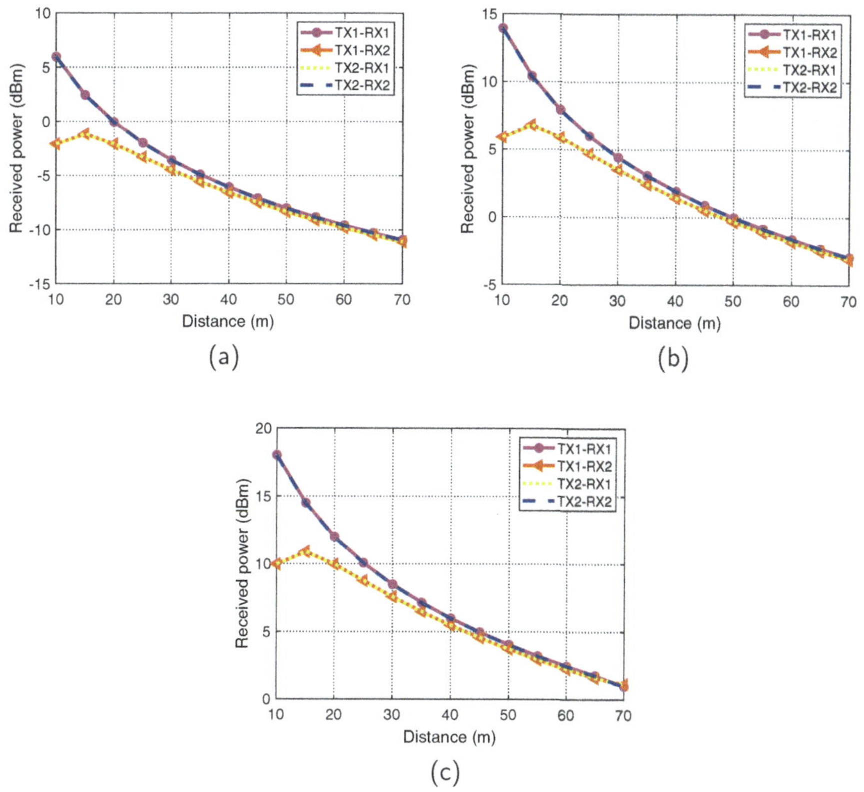

Fig. 3 Received power versus distance for: (**a**) $Dr = 1\,\text{cm}$ (**b**) $Dr = 2.5\,\text{cm}$ (**c**) $Dr = 4\,\text{cm}$ for $d_h = 0$ and clear weather

and thick foggy weather, respectively, which indices a degradation of only 0.4 dBm and 0.5 dBm compared to clear weather. On the other hand, for $d_x = 70\,\text{m}$, the received power is -11 dBm for clear weather. This further reduces to -13 dBm and -15 dBm for moderate and thick foggy weather, respectively, with a relative difference of -2 dBm and -4 dBm with respect to clear weather.

In Fig. 5, we present the BER versus the Receiver SNR for SMP with different lateral shifts while keeping the $D_r = 1\,\text{cm}$ (the minimum aperture) and the clear weather. It is observed that the lateral shift has a significant impact on the BER performance where the larger d_h the lower BER. This is due to the fact that in the perfect alignment scenario, the channel gains are quite similar, providing high channel correlation while the reverse is correct for lateral shifts. For example, consider $d_x = 25\,\text{m}$. The required SNR to achieve BER of $\leq 10^{-3}$ at $d_h = 0\,\text{m}$ is 35 dB as shown in Fig. 5a. This reduces to 32 dB and 30 dB for $d_h = 1\,\text{m}$ and $d_h = 2\,\text{m}$, respectively (see Fig. 5b,c). It is also observed that the propagation distance (d_x) has a noticeable effect on the BER performance. For example, consider

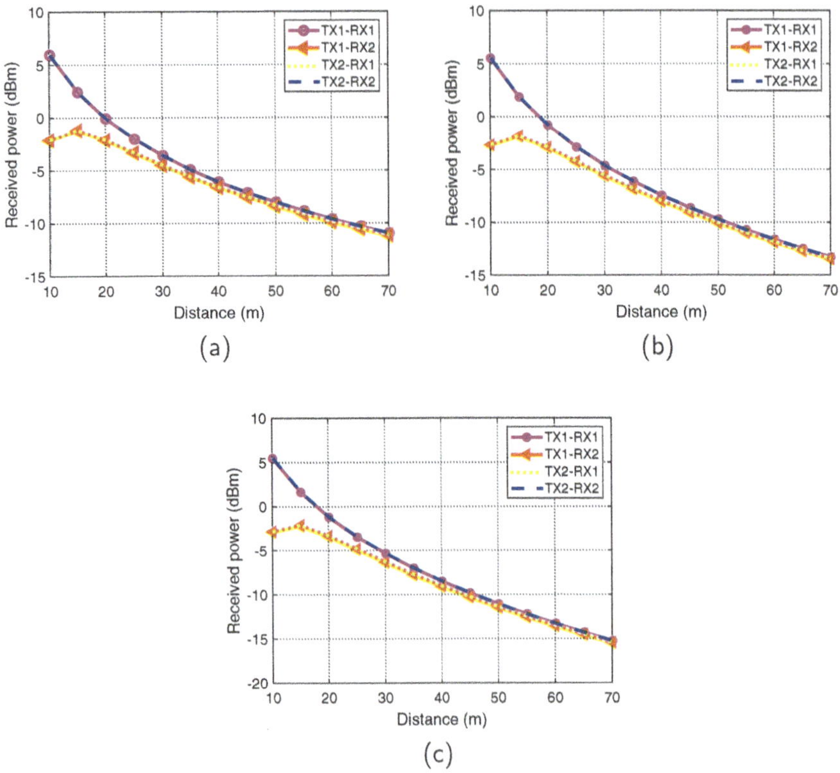

Fig. 4 Received power versus distance for: (**a**) Clear weather. (**b**) Moderate foggy weather. (**c**) Thick foggy weather for $d_h = 0$ m and $Dr = 1$ cm

$d_h = 0$ m. The required SNR value to achieve BER of $\leq 10^{-3}$ is 35 dB for $d_x = 25$ m. This significantly increases to 47 dB and 52 dB for $d_x = 50$ m, and $d_x = 70$ m, respectively.

5 Conclusion

In this paper, we have investigated the performance of a 2 × 2 MIMO VLC based-V2V system. Two HLs of the source vehicle are deployed as wireless transmitters while two PDs installed at the destination vehicle are served as wireless receivers. A most recent realistic path loss model has been utilized which is based on the non-sequential ray-tracing for channel modeling. Both vertical and horizontal displacements same as weather conditions and the receiver aperture have been taken into account. Our results reveal that the lateral shift between cars has a significant effect on the received power and BER performance, particularly at shorter distances.

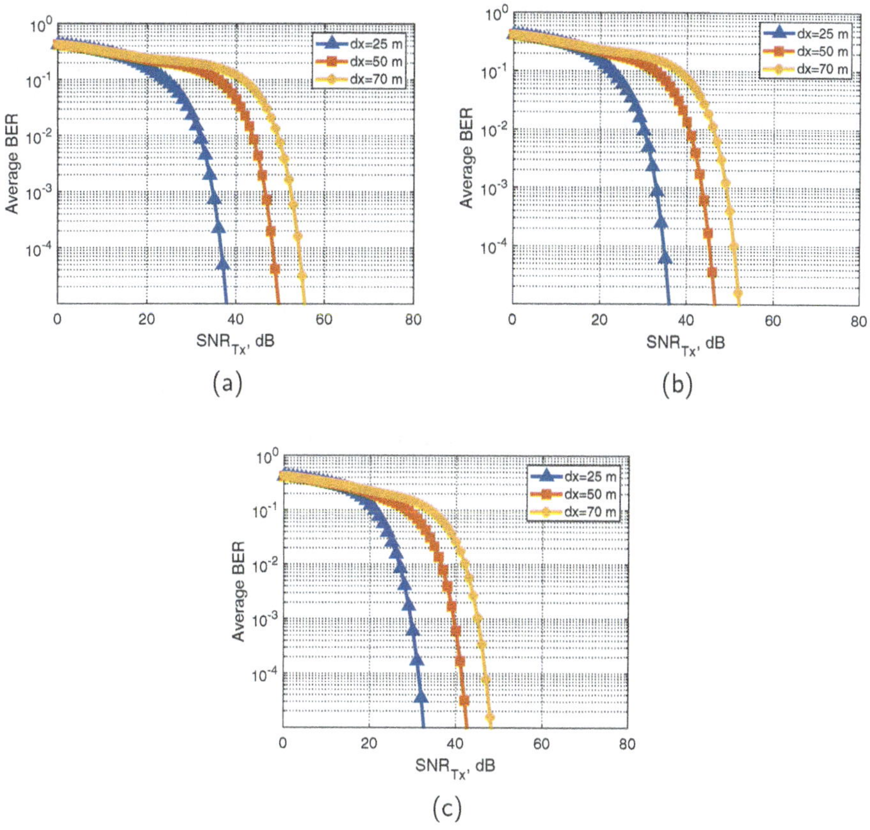

Fig. 5 BER versus Receiver SNR for: (**a**) $d_h = 0\,\mathrm{m}$ (**b**) $d_h = 1\,\mathrm{m}$ (**c**) $d_h = 2\,\mathrm{m}$, for $Dr = 1\,\mathrm{cm}$ and clear weather

The results demonstrate also that the system performance is degraded when the longitudinal distance becomes large or when the atmosphere becomes more cloudy. Furthermore, the results demonstrate the ability of parallel data transmission for V2V systems based on VLC technology. This will enlarge the achievable data rate and could be investigated in our future works.

References

1. Alam, M., Rayes, A., He, X., Atiquzzaman, M., Lloret, J., Tsang, K.F.: Guest editorial introduction to the special issue on dependable wireless vehicular communications for intelligent transportation systems (ITS). IEEE Trans. Intell. Transp. Syst. **19**(3), 949–952 (2018)
2. Perallos, A., Hernandez-Jayo, U., Onieva, E., García Zuazola, I.J.: Intelligent Transport Systems: Technologies and Applications. John Wiley & Sons, Hoboken (2015)

3. Gyawali, S., Xu, S., Qian, Y., Hu, R.Q.: Challenges and solutions for cellular based v2x communications. IEEE Commun. Surv. Tutorials **23**, 222–255 (2020)
4. Wu, X., Subramanian, S., Guha, R., White, R.G., Li, J., Lu, K.W., Bucceri, A., Zhang, T.: Vehicular communications using DSRC: challenges, enhancements, and evolution. IEEE J. Sel. Areas Commun. **31**(9), 399–408 (2013)
5. Abboud, K., Omar, H.A., Zhuang, W.: Interworking of DSRC and cellular network technologies for V2X communications: a survey. IEEE Trans. Veh. Technol. **65**(12), 9457–9470 (2016)
6. Vivek, N., Srikanth, S.V., Saurabh, P., Vamsi, T.P., Raju, K.: On field performance analysis of IEEE 802.11 p and wave protocol stack for v2v & v2i communication. In: International Conference on Information Communication and Embedded Systems (ICICES2014), pp. 1–6. IEEE, Piscataway (2014)
7. Jovicic, A., Li, J., Richardson, T.: Visible light communication: opportunities, challenges and the path to market. IEEE Commun. Mag. **51**(12), 26–32 (2013)
8. Uysal, M., Ghassemlooy, Z., Bekkali, A., Kadri, A., Menouar, H.: Visible light communication for vehicular networking: performance study of a v2v system using a measured headlamp beam pattern model. IEEE Veh. Technol. Mag. **10**(4), 45–53 (2015)
9. Eldeeb, H.B., Sait, S.M., Uysal, M.: Visible light communication for connected vehicles: how to achieve the omnidirectional coverage? IEEE Access **9**, 103885–103905 (2021)
10. Căilean, A.-M., Dimian, M.: Current challenges for visible light communications usage in vehicle applications: a survey. IEEE Commun. Surv. Tutorials **19**(4), 2681–2703 (2017)
11. Alsalami, F.M., Aigoro, N., Mahmoud, A.A., Ahmad, Z., Haigh, P.A., Haas, O.C.L., Rajbhandari, S.: Impact of vehicle headlights radiation pattern on dynamic vehicular VLC channel. J. Lightwave Technol. **39**(10), 3162–3168 (2021)
12. Cui, Z., Wang, C., Tsai, H.-M.: Characterizing channel fading in vehicular visible light communications with video data. In: 2014 IEEE Vehicular Networking Conference (VNC), pp. 226–229. IEEE, Piscataway (2014)
13. Avătămăniţei, S.-A., Căilean, A.-M., Zadobrischi, E., Done, A., Dimian, M., Popa, V.: Intensive testing of infrastructure-to-vehicle visible light communications in real outdoor scenario: evaluation of a 50 meters link in direct sun exposure. In: 2019 Global LIFI Congress (GLC), pp. 1–5. IEEE, Piscataway (2019)
14. Kim, Y.H., Cahyadi, W.A., Chung, Y.H.: Experimental demonstration of VLC-based vehicle-to-vehicle communications under fog conditions. IEEE Photonics J. **7**(6), 1–9 (2015)
15. Memedi, A., Tsai, H.-M., Dressler, F.: Impact of realistic light radiation pattern on vehicular visible light communication. In: GLOBECOM 2017 - 2017 IEEE Global Communications Conference, pp. 1–6 (2017)
16. Eldeeb, H.B., Eso, E., Uysal, M., Ghassemlooy, Z., Zvanovec, S., Sathian, J.: Vehicular visible light communications: the impact of taillight radiation pattern. In: 2020 IEEE Photonics Conference (IPC), pp. 1–2. IEEE, Piscataway (2020)
17. Viriyasitavat, W., Yu, S.-H., Tsai, H.-M.: Short paper: channel model for visible light communications using off-the-shelf scooter taillight. In: 2013 IEEE Vehicular Networking Conference, pp. 170–173 (2013)
18. Eldeeb, H.B., Miramirkhani, F., Uysal, M.: A path loss model for vehicle-to-vehicle visible light communications. In: 2019 15th International Conference on Telecommunications (ConTEL), pp. 1–5. IEEE, Piscataway (2019)
19. Eldeeb, H.B., Eso, E., Jarchlo, E.A., Zvanovec, S., Uysal, M., Ghassemlooy, Z., Sathian, J.: Vehicular VLC: a ray tracing study based on measured radiation patterns of commercial taillights. IEEE Photon. Technol. Lett. **33**(16), 904–907 (2021)
20. Liu, J., Chan, P.W.C., Ng, D.W.K., Lo, E.S., Shimamoto, S.: Hybrid visible light communications in intelligent transportation systems with position based services. In: 2012 IEEE Globecom Workshops, pp. 1254–1259. IEEE, Piscataway (2012)
21. Cailean, A., Cagneau, B., Chassagne, L., Topsu, S., Alayli, Y., Blosseville, J.-M.: Visible light communications: application to cooperation between vehicles and road infrastructures. In: 2012 IEEE Intelligent Vehicles Symposium, pp. 1055–1059. IEEE, Piscataway (2012)

22. Aly, B., Elamassie, M., Eldeeb, H.B., Uysal, M.: Experimental investigation of lens combinations on the performance of vehicular VLC. In: 2020 12th International Symposium on Communication Systems, Networks and Digital Signal Processing (CSNDSP), pp. 1–5 (2020)
23. Bao, X., Zhu, X., Song, T., Ou, Y.: Protocol design and capacity analysis in hybrid network of visible light communication and OFDMA systems. IEEE Trans. Veh. Technol. **63**(4), 1770–1778 (2013)
24. Demir, M.S., Eldeeb, H.B., Uysal, M.: CoMP-based dynamic handover for vehicular VLC networks. IEEE Commun. Lett. **24**(9), 2024–2028 (2020)
25. Bazzi, A., Masini, B.M., Zanella, A., Calisti, A.: Visible light communications as a complementary technology for the internet of vehicles. Comput. Commun. **93**, 39–51 (2016)
26. Eldeeb, H.B., Yanmaz, E., Uysal, M.: MAC layer performance of multi-hop vehicular VLC networks with CSMA/CA. In: 2020 12th International Symposium on Communication Systems, Networks and Digital Signal Processing (CSNDSP), pp. 1–6 (2020)
27. Yahia, S., Meraihi, Y., Gabis, A.B., Ramdane-Cherif, A.: Multi-directional vehicle-to-vehicle visible light communication with angular diversity technology. In: 2020 2nd International Workshop on Human-Centric Smart Environments for Health and Well-being (IHSH), pp. 160–164. IEEE, Piscataway (2021)
28. Eldeeb, H.B., Uysal, M.: Vehicle-to-vehicle visible light communication: how to select receiver locations for optimal performance? In: 11th International Conference on Electrical and Electronics Engineering (ELECO), pp. 402–405. IEEE, Piscataway (2019)
29. Petrariu, A.I., Lavric, A., Coca, E.: VLC for vehicular communications: a multiple input multiple output (MIMO) approach. In: 2018 International Conference on Development and Application Systems (DAS), pp. 134–137. IEEE, Piscataway (2018)
30. Luo, P., Ghassemlooy, Z., Minh, H.L., Bentley, E., Burton, A., Tang, X.: Bit-error-rate performance of a car-to-car VLC system using 2 × 2 MIMO. Mediterr. J. Comput. Netw. **11**, 400–407 (2015)
31. Farahneh, H., Kamruzzaman, S.M., Fernando, X.: Differential receiver as a denoising scheme to improve the performance of v2v-vlc systems. In: 2018 IEEE International Conference on Communications Workshops (ICC Workshops), pp. 1–6. IEEE, Piscataway (2018)
32. Liu, W., He, X.: Performance analysis of MIMO visible light based v2v communications. In: 2019 IEEE 89th Vehicular Technology Conference (VTC2019-Spring), pp. 1–4. IEEE, Piscataway (2019)
33. Gupta, V., Singhal, R.: Performance analysis of a visible light vehicle-to-vehicle wireless communication system. In: 2019 TEQIP III Sponsored International Conference on Microwave Integrated Circuits, Photonics and Wireless Networks (IMICPW), pp. 521–523. IEEE, Piscataway (2019)
34. Eldeeb, H.B., Mana, S.M., Jungnickel, V., Hellwig, P., Hilt, J., Uysal, M.: Distributed MIMO for Li-Fi: channel measurements, ray tracing and throughput analysis. IEEE Photon. Technol. Lett. **33**(16), 916–919 (2021)
35. Khalid, A., Asif, H.M., Mumtaz, S., Al Otaibi, S., Konstantin, K.: Design of MIMO-visible light communication transceiver using maximum rank distance codes. IEEE Access **7**, 89128–89140 (2019)
36. Karbalayghareh, M., Miramirkhani, F., Eldeeb, H.B., Kizilirmak, R.C., Sait, S.M., Uysal, M.: Channel modelling and performance limits of vehicular visible light communication systems. IEEE Trans. Veh. Technol. **69**(7), 6891–6901 (2020)
37. Mmbaga, P.F., Thompson, J., Haas, H.: Performance analysis of indoor diffuse VLC MIMO channels using angular diversity detectors. J. Lightwave Technol. **34**(4), 1254–1266 (2015)

Efficient Big Data Architecture Based on Micro Service

Conversion into Structured Record

Fabrice Mourlin, Cyril Dumont, and Laurent Nel

1 Introduction

The use of Big Data has passed its first decade, and applications have followed the evolution of software. Developers have moved from Hadoop Map/Reduce development to higher-level frameworks like Spark, or Flink. Computing cases are executed on dedicated clusters, and their management is often global for all applications. This approach introduces difficulties. They appear first when the calculation cases are run on the clusters. In order to keep the applications from being too large, the deliverables must not contain the libraries they use but share the libraries installed on the cluster. This forces the applications to share the same framework versions. The example of Cloudera/Hortonworks virtual machines is an example of a software stack that is perfectly configured so that all the software installed are compatible with each other.

F. Mourlin (✉)
Univ Paris Est Creteil, LACL, Creteil, France
e-mail: fabrice.mourlin@u-pec.fr

C. Dumont
Leuville Objects, Versailles, France
e-mail: cyril.dumont@leuville.com

L. Nel
Leuville Objects, Versailles, France

Univ Paris-Saclay, Polytech Paris-Saclay, Orsay, France
e-mail: laurent.nel@leuville.com

M. D. Hina et al. (eds.), *Advances in Computational Intelligence and Communication*, EAI/Springer Innovations in Communication and Computing,
https://doi.org/10.1007/978-3-031-19523-5_5

Thus, in the version of the virtual machine HDP 2.6.5, all applications using Spark must take version 2.3.0, even if some would like to upgrade their application. On the other hand, the change of version of the virtual machine leads to a change in the whole software stack that all applications support. Consequently, it is not easy for a technical architect to know where to place the cursor between a ready-made VM or an adhoc solution. It seems natural today to want to decouple the management of applications. Other aspects come into play such as the management of computing resources. In the case of YARN or Spark, how to manage the sequencing of jobs. If previously this sequencing was centralized, should we distribute this sequencing by exploiting the isolation of jobs? The evolution of software needs to bring out software properties that must be taken into account: ensure portability between two clusters, simplify applications in order to focus on the business, and remove the notion of resource silo to better share resources.

If we consider a major change such as the move to Hadoop 3 two years ago. Its implementation took a considerable amount of time compared to the actual technical changes. Indeed, this strategic decision brings greater availability to our data lake, but it also means updating all the applications that access our distributed file system. From these changes come the time costs and sometimes the blockages to implementing important updates of an architecture too monolithic.

Our experience with Big Data projects highlights another limitation. Over the past decade, some software in the HDP stack has fallen into disuse, while others have evolved and still meet the needs of projects. This is the case of Spark, Cassandra, among others. However, the software stack in use has so many interoperating software components that it has become very difficult to consider upgrades. In addition, new versions of this software stack are rare, while the need for software evolutions is present. In hindsight, we can see that the calculation components and the storage components are strongly coupled. It is therefore essential to involve an IT expert to configure the data lake with regard to the volume of data to be processed in order to avoid blockages when accessing the data during analysis.

It is easy to realize that the management of monolithic clusters has introduced management costs that medium-sized companies such as mutual insurance companies, transport companies, etc., whose turnover does not allow for cost overruns due to the computer system, cannot bear.

Initially, the main components of the Hadoop software stack were written in Java. However, the field of analysis and learning has very well-known frameworks in Python. Of course, it is possible to use Python scripts to do Map/Reduce in Hadoop, and it is easy to use the PySpark library in Python, but the use of Python frameworks such as Panda or Koala in the Hadoop stack is a computing feat. There is even a subtomorrow of AI where Python frameworks are more referenced: it is deep learning. An AI expert will want to use PyTorch developed by Facebook. The integration of this framework in the Hadoop stack remains to be done for each cluster used. We note that interoperability in data access is a property that Big Data developers want.

The work presented in this paper represents our choice to evolve the Big Data architecture toward a solution that is more open to project-specific technology choices but also to a looser coupling between computational resources, sequencing, and storage resources. To this end, we have structured the rest of the paper as follows: in the next section, we place our work in the context of Big Data architectures. The third section describes our use case supporting this study. In the fourth section, we detail our technical architecture and highlight the software properties we are looking for. Finally, in the last section, we provide metrics to validate our choices and then conclude on the contributions of this work for new projects.

2 Related Works

2.1 Contact of An Essential Platform

The work on architectures for Big Data is numerous and reflects the evolution not only of computing needs on the one hand but also of data consumption. In [1], M. Bilal proposes a Big Data architecture for construction waste analytics. They identify critical components of the proposed Big Data-based waste analytics architecture. This includes specific graph-oriented database and specific programming library. This solution not only allows storing large data sets but also provides a full Java solution compatible with the Spark framework used in this work.

The work of F. Matsebula and all [2] concerns data acquisition for student orientation at the university. In this context, the authors have defined a dedicated storage system for the volume and nature of data received. In particular, the use of temporal data adds a technical difficulty because the choice of Hadoop Database (HBase) results in encoding all information in binary. Respecting the standard HDP software stack and information routing tools such as the tool called sqoop introduces costs at runtime but also heaviness when updating a part of the software to use Kafka queues for data routing.

K.M.L Jones has done work on the same subject [3] but with an interest in the traceability of information. Who uses the data from the data lake and especially the trace of the results of the analyses made? Thus, a platform named P3P for Platform for Privacy Preferences provides a dashboard so that the students control and validate the use of their private data. He refers to other works where semantic information is added for profile definition. He then proposes an implementation via Apache Ranger that allows a global approach to security for Hadoop clusters. This tool acts as a centralized platform to define, administer, and manage security policies in a consistent manner across the Hadoop ecosystem. In this case, it is the only solution on the Cloudera platform, and its choice is almost mandatory.

J. Wang wrote a survey with several co-authors where he points out that architects consider a Big Data platform more as an interconnected software block than as a scalable software platform. In his presentation, each tool has a role that explains its presence in the platform [4]. He points out that these preconfigured virtual machines are made to help the novice to get up to speed with Big Data technologies. He concludes that after several experiences with such tools, it is advisable to review one's knowledge in order to define a Big Data specification where only useful tools will be kept. The goal is to simplify an architecture so that it becomes more maintainable.

2.2 Research on the Evolution of Architectures

Evolutions have been made to analyze business data. In the context of process monitoring, M.F. Huber processes a large amount of complex events using Spark job programming [5]. Spark CEP is a stream processing engine on top of Apache Spark supporting continuous query language. In this use case, the implementation of a data lake is based on Apache Cassandra, a column-oriented database. This work shows that it is possible to change the architecture, but this requires recreating a new monolithic solution from the persistence layer. The choice of Cassandra instead of HBase is explained by the structured nature of the stored data. Typing is more important when analyzing and building predictive models. M. Y. Santos uses the Hadoop stack in his study and insists on the constraints imposed in particular on the Spark streaming versions and more specifically the Java 8 language [6]. The information coming from the devices are routed with REST services. The authors have chosen Cassandra as a NoSQL database whose version is linked to that of Spark streaming. An update of Java to an even slightly higher version is not possible.

It is obvious that any assembly of strongly coupled software is not scalable. It is necessary to turn to new software architectures. T Ivanov proposes an abstract architecture in the manner of the Model-Driven Architecture (MDA) description [7]. However, this is only an anonymization of a current architecture with still strong dependencies between components. M. S. Hadj Sassi has published about a work on an architecture for the collection of data from sensors and their analysis [8]. He focuses his solution on the definition of business services specific to the distribution of sensors, and he adopts above all an asynchronous coupling between services. Throughout the paper, he presents a business workflow based on this principle where each phase is decomposed into a component diagram illustrating the asynchronous points.

G. Moktari followed the same approach for the definition of a layered architecture in the context of Big Data applications for the smart home [9]. The author insists mainly on the real-time aspects even if some asynchronous points are put forward, for example, at the end of the data collection. As the paper describes a case study where the information comes from buildings with sensors, the data collections are done via a cloud where the Big Data processing takes place. The use of REST APIs indicates that the real-time aspects are not preserved, but the layered architecture underlines a geographical division of the processing and necessarily the isolation of some software. Through these works, we can see software properties that have become important because Big Data technologies are evolving rapidly. In addition, the basics of use have also become more complex with an increasing amount of data and access requirements that need to be continuously improved.

3 Use Case Description

The description of our use case is based on our desire to monitor the activity of our information system. This includes several application servers and data management servers, interconnected by a software bus. It enables intelligent message routing between applications and provides a first level of fault tolerance in the event of a service failure.

Our servers provide log files, but also our applications are deployed on the servers. Many formatted files are thus written in different directories. To perform a centralized log analysis, a preparatory step consists in moving the files to a dedicated machine. A second step consists in analyzing the data to keep the useful parts on the one hand and to index the key parts on the other hand. This pipeline continues with the use of a statistical model to predict the actions to be planned (Fig. 1). Finally, the last step concerns the collection of metrics in order to evaluate the monitoring process.

During our first prototypes, the volume of data processed exceeded 25 MB per hour, and it became evident that such a sequential process could not meet our needs. The choice of a Big Data cluster for the processing of such volumes of text is legitimate, especially since this work relates to the monitoring of distributed systems.

Each of these steps is implemented using a containerized micro service. The exchanges between steps use message queues.

3.1 Collection of Log Data

The data collection step consists of moving the log files from the application server's configuration directory to a Kafka topic. This Message-Oriented Middleware (MOM) has the role of filling the message brokers. Thus, the application server or message producer publishes its message in a topic, while the consumers subscribe to this topic to be able to read the messages to clean them up. The Publish/Subscribe

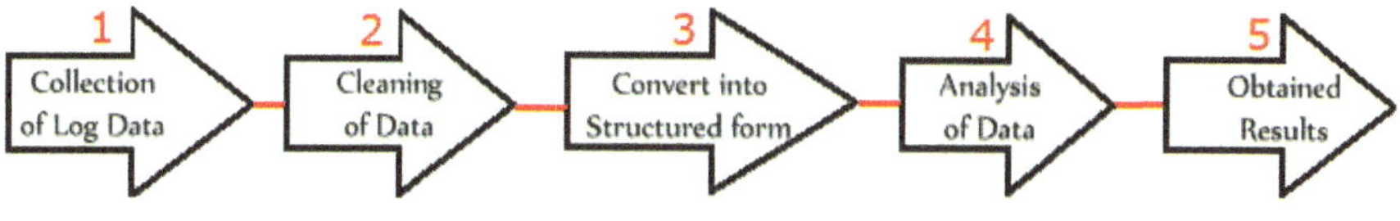

Fig. 1 Log analysis process: Big Data pipeline with five steps

strategy is the simplest way to ensure scaling. The Read Once Only Once principle is implemented in Apache Kafka. A message is read only once by a single consumer. Topics are divided into several partitions to ensure high read and write throughput. The number of partitions is configurable for each topic. The greater the need for speed in writing and reading, the higher the number of partitions per topic. Each partition is replicated on several servers to obtain good resistance to failures and ensure a high availability operation. Each message sent in a topic having a fixed lifetime, the cleaning step must be operational during this lifetime to have access to the message.

3.2 Cleaning of Data

This process consists of modifying the dataset in order to modify or delete incomplete, irrelevant, corrupted, duplicated, or badly formatted data. Indeed, when collecting data from source servers, data can quickly become mislabeled or duplicated within the same set. In this case, the service at this stage modifies or deletes the data from the initial set, as it is generally neither necessary nor useful in the analysis process. Even worse, they may distort the results and affect the accuracy of the results. The quality of the results depends on the quality of the data. In order to improve this step, all application servers are configured so that their logs respect a format. This format is not unique because it is not currently possible to impose this constraint, but the format provided by the configuration is sufficient to validate that all data is present. Moreover, conversions are made so that the time units are easily comparable. In the output, we have uniform datasets, easier to process for storage in our data lab.

3.3 Convert into Structured Record

Many works already exist on text mining. R. Jundal has even proposed a framework for the conversion of unstructured data into structured data [10]. It analyzes the text from different points of view (lexical, syntactic, and semantic) to produce an intermediate or enriched representation with weights to highlight the terms. We have adopted a comparable approach although this step aims at storing the data in a document-oriented NoSQL database (MongoDB). We exploit our initial log formats to preserve the knowledge from our data. We also store our semantic metrics based on term occurrences. For example, we can easily extract a timestamp from an event or a user ID or an IP address, etc. More interestingly, we can search for the most numerous alerts concerning an IP address, or those concerning a user identifier over a time interval. The query language is rich and supports text search, aggregation features, and CRUD operations. Additional tools such as Apache Hive can easily be configured to query data using traditional SQL.

3.4 *Analysis of Data*

From the data stored in the database, it is then possible to build a statistical model that allows us to make predictions. In our case, we have chosen a linear support vector machines (SVM) model. This AI algorithm has implementations in many libraries, itself written in different languages. It allows us to rank the probability that words in a line of logs are correlated with an incident. For example, the log level error corresponds to a serious incident, and its higher probability leads to a request for maintenance operation on the server whose IP address is known. This supervised machine learning algorithm requires many data in order to have a good accuracy. This leads to a high cost of implementation because the training phase is long. One of the main difficulties is the detection of rare events because any instance of an event in the logs is made of different data. A second difficulty appears when the application servers themselves evolve. When the deployed applications change version or simply when new applications are deployed, it is possible to notice a gap between the training data and the analyzed data. In this case, we set up metrics to trigger a re-learning phase or context update. The two main metrics concern the counting of new events, new identifiers, and unknown formats.

3.5 *Obtained Results*

We attach to the use of an AI model a report to provide activity monitoring. Not only does it describe the activity periods but also the volume of requests processed, and the maintenance operation triggers. It also provides the update of the reinforcement metrics and a description of the input datasets. The goal of these results is to provide useful information to validate or invalidate the proper use of our AI model. This information is semi-structured and can be used by a report generation tool such as Jasper Report.

4 Technical Architecture

Our technical choices are based on classic specifications for a log analysis application. However, with our experience and our desire to make our project evolve at our convenience, we have chosen to isolate each technical aspect in an application container. Thus, we have additional means for the administration of our distributed system. The software architecture diagram in Fig. 2 illustrates a first level of architecture, where each UML component is concretized by a Docker image to be instantiated by an orchestration tool (Swarm in local or prototyping or Kubernetes in cluster mode).

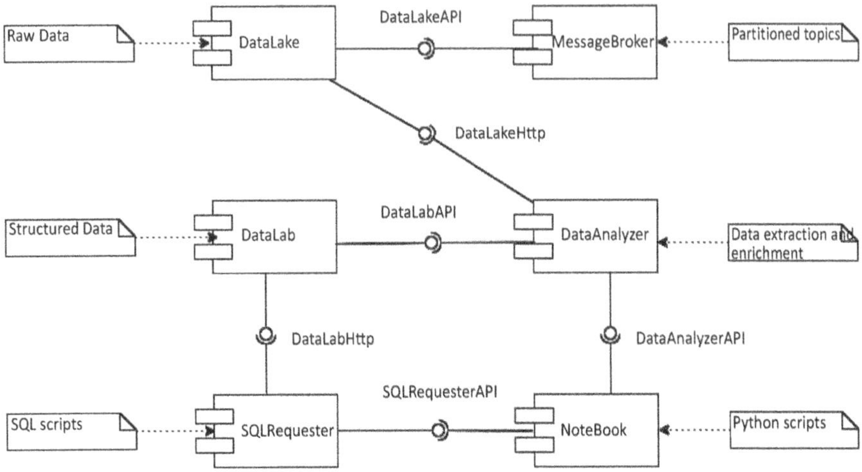

Fig. 2 A deployment diagram of our Big Data orchestration

One of the first steps in defining our architecture was to define the technical perimeter we needed. To keep things simple, we proceeded by successive layers: first, a data lake to receive the data to be analyzed from its production source. Then, we defined a data lab to store structured data ready to be used for our analyses. More precisely, we selected HDFS, Hive, Spark, MongoDB, Zeppelin, Kafka, Zookeeper, and Streamsets, a sufficient subset of tools to build our workflow. To isolate the impact of each tool, we built a specific Docker image to expose resources (volumes, network, ports, variables, etc.). We use Kubernetes to orchestrate our containers.

4.1 *The Message Broker*

We chose a distributed tool for streaming data: Apache Kafka because it allows us to publish, store, process, and subscribe to recording streams in real time. Moreover, it is designed to manage data streams coming from several sources (our application servers) and provide them to our data lake while respecting the subdirectory breakdown by log type. In addition to moving data to our data lake, we configure a Kafka service to clean up the log data before depositing it in an agreed-upon data lake directory. We have thus implemented an asynchronous integration of the log data. This implies data replication in the data lake increases accessibility by Big Data analyzers.

To run Kafka, we also need to run Zookeeper. Even though it is possible to run Zookeeper, we preferred that each container be responsible for only one component. The tricky part of running the official Kafka images is to set the environment variables correctly, especially for the choice of static IP. We defined the

Zookeeper and Kafka services as belonging to the same bridge network as the other components. In Docker terms, a bridge network uses a software bridge that allows containers connected to the same bridge network to communicate while providing isolation from containers that are not connected to that bridge network. In addition, to ensure monitoring, we have exposed the standard ports of these tools.

4.2 A Data Lake

Our data lake is our storage repository that stores a large amount of log data files. This is the phase to store every type of log data in its native format with no fixed limits on size. We have defined large containers, which rely on HDFS services. This means the definition of two main services, name node and data node, which are based on official Docker image. Each service has its own volume and a set of environment variables. The service port is mapped to public ports, and they belong to a common network. The instances of the name node and the data nodes are deployed over a network with predefined IP addresses. The whole data lake has sub-directories per file format. Moreover, depending on the subdirectory, we have applied cleaning routing on the log data. The DataLake component implements two interfaces in Python; on the one hand, the *DataLakeAPI* interface offers the operations of depositing log data in the data lake, that is, in the appropriate directory depending on the initial format. This interface allows the acknowledgment of the deposited data (event timestamp, size, etc.). The other interface *DataLakeHttp* allows the extraction of data by the *DataAnalyser* component. In the same way, any extraction operation is traced, and the event log via this interface is readable.

We do not run Yarn (Yet Another Resource Negotiator) but only HDFS (Hadoop Distributed File System). For our Spark jobs, we only use stand-alone management. To run HDFS, we use an official image. The entire configuration is defined in the hadoop-hive.env file (replication rate, directories, etc.). We have created a volume to store the data. Therefore, we do not delete the data in these mounted folders when the Docker containers are stopped or restarted.

4.3 A Data Analyzer

The Data Analyzer component is based on an Apache Spark image; it includes Python scripts. Their main role is to extract log data from the data lake in order to validate each document to store it in the data lab. For this purpose, we use the formats associated with the sources of each log. Then, the data is standardized to convert if necessary some values such as dates, times, or network address. Finally, the data put in JSON format are exported to the data lab, which centralizes the data to take advantage of this data, to transform it into benefit. This component uses two interfaces, on the one hand, the *DataLakeHttp* interface, which is an overlay of the HDFS REST API; essentially, these are log data reading and control operations (access to file dates, marking of read files, etc). On the other hand, the *DataLabAPI* interface, which offers import operations of formatted data into the persistence system and acknowledgments of these operations. The log of the data lab operations is also accessible in read mode.

The definition of this component includes several services: on one hand, a spark master and on the other hand, spark workers. From official Docker images for these services adapted to Hadoop 2.7, we have configured this federation by using environment variables so that these services can interact with the data lake, which is based on an HDFS layer and the SQL Requester, which is based on a Hive instance. Again, all these services belong to the same bridge network, and the monitoring addresses are exposed for the administration of the use case.

4.4 The Data Lab

We have created a data lab for the implementation of a persistence system to standardize the use of log data. This component represents the way to use log data in particular to build our AI models from tables. For this purpose, we have configured the persistence services of MongoDB via the use of the official image. This involves defining the data schema used by our Spark applications. As for the other components, we have exposed the ports useful for software monitoring and defined the membership of these instances to the same bridge network.

In MongoDB, data are modeled as a document in a JSON style. We do not talk about tables or records anymore but about collections and documents. This data management system avoids the need to join tables because the entire information specific to a certain data is stored in the same document. Among the Mongo collections created, one contains the standardized information from the logs; another one contains semantic information on the data of the first collection such as the weight of a term in the document and the weight of a term in the collection. These notions are crucial when searching with the Mongo API. The *DataLab* component exposes two interfaces implemented in Python via the PyMongo library. On the one hand, the *DataLabAPI* interface performs the import of log data and their tracking; on the other hand, the *DataLabHttp* interface supports querying on log data and associated metrics. We have developed these operations so that the evaluation of queries is driven by the data lab and not by the underlying Mongo DB system. Our goal is not to search for exact data known beforehand but to search for data related to an upcoming incident. Hence, the computation of a distance (named *logdis*) between the query and the documents in our collection. This results in real weighting, which is used to rank the results.

4.5 A SQL Requester

The *SQLRequester* component is created to satisfy the demand of users who wish to perform SQL queries although the data lab does not support it by default. It relies on the use of an instance of the Hive tool and a local database to save the metadata. We have created external tables like our collections in the data lab. To ensure compatibility, these tables represent the worst case of the imported documents. Then, the Hive tool performs the transcription of any extraction requests to the data lab. To save the metadata of Hive, you use a Postgresql Docker image of external database not shown in Fig. 2. The role of this external database is to keep all the data and metastore useful to Hive and thanks to Docker volumes, mount a data folder somewhere in a local machine on the network. The *SQLRequester* component uses the *DataLabAPI* interface to evaluate SQL scripts written by an end user or included in a note built from a prototyping tool. This component provides a Python implementation of the *SQLRequestAPI* that provides a specialized REST API for accessing structured data from our Data lab from a notebook.

This orchestration is composed of three main services. The Winter Server service is built from a Hive image for access to a Postgresql database. This service depends on the Hive Metastore service, which is also based on an image having access to the Postgresql server. Finally, a persistence service is based on a Postgresql Server image. All the services are as before on the same network with the exposure of the ports attached to the monitoring.

4.6 The NoteBook

This component provides a tool for writing computer code and sharing it to collaborate on a team project. The sharing methods are done via script exchange in JSON format. The notebook chosen is Zeppelin for its simplicity and the versatility of the scripting languages offered. It provides a first level of visualization comparable to what some office tools do. It also accepts the use of graphic libraries. In this study, we built PySpark scripts to construct an SVM model using the Spark MLlib library. Spark MLlib offers an implementation of support vector machines (SVM) with a linear kernel (suitable for linearly separable data). In machine learning, support vector machines (SVMs) are supervised learning models with associated learning algorithms that analyze data and recognize patterns, used for classification analysis. Given a set of log training examples, each belonging to one of the few class labels, an SVM algorithm builds a model that assigns new examples to one label or another. Unlike logistic regression, SVM is a non-probabilistic binary linear classifier. An SVM model is a representation of the log document examples as points in space, mapped so that the log examples of the separate categories are divided by a clear gap that is as wide as possible. New log examples are then mapped

Table 1 SVM model trainer based on our log data

```
logModel = SVMWithSGD.train(logData,~iterations=100,
           step=1.0, regParam=0.01,
           miniBatchFraction=1.0,
           initialWeights=None, regType="l2",
           intercept=False, validateData=True,
           convergenceTol=0.001)
```

into that same space and predicted to belong to a category based on which side of the gap they fall on.

For the training step, Spark Mllib encapsulates SVMWithSGD support vector machine classifier, which .train () method of training model, call the following form:

The parameters are as follows:

- *logData*: The training input log data, the data format of format *LabeledPoint*
- *iterations*: Use SGD iterations, default is 100
- *step*: SGD each execution iteration step size, the default is 1
- *miniBatchFraction*: Small quantities of stochastic gradient descent method participation rate per sample calculation, values from 0 to 1, the default is 1
- *initialWeights*: Initialization factor, the default is None
- *regParam*: Regularization coefficient size
- *regType*: Regularization type "l1" or "l2" or "None," the default is "l2,"

This part of the Python script (Table 1) shows that the use of a notebook has become essential in the world of data science but does not have the functionality of a development environment. The Notebook component is based on an image of Spark Zeppelin, its basic interpreter is Spark engine. In addition, it is essential to provide the url of the Spark Master, as well as the URL of the Data lake NameNode and the Hive Server. It also has its own volume to keep the scripts and load them on demand. Of course, this component belongs to the same Docker bridge network.

5 Results and Measures

5.1 Runtime Configuration

We obtained our results by running Python scripts from our NoteBook component. In addition, the whole orchestration must be executed on our Big Data cluster. Kubernetes is a container orchestration engine, which ensures there is always a high availability of resources. Apart from that, it also has the following features: self-healing, automatic rolling updates and rollback, resource management, service discovery, load balancing, and service discovery.

A Kubernetes architecture has two major component type; they are master components and worker components. The master components treat all the requests from the user using API, and *kubectl* are sent to the master component, that is the API Server. The worker components come under the nodes

From the NoteBook, we use spark-a *submit command* directly to submit our Spark application to a Kubernetes cluster. Once submitted, the following events occur:

- Creation of a Spark driver running as a Kubernetes pod
- Creation of executors, which are also run within Kubernetes pods, connect to them, and execute the application code
- Termination and cleanup of executor pods occur when the application completes However, the driver pod persists, logs, and remains in "completed" state in the Kubernetes API until it's eventually garbage collected or manually cleaned up.

5.2 Model Assessment

The use of forecast accuracy as an indicator to assess the model for simplicity, custom function calculation accuracy rate (Table 2). Our result for accuracy is 0.747573914867.

We proceeded to the evaluations of the main parameters of the model by specifying ad hoc functions: evaluation parameters iterations, parameter evaluation step, evaluation parameters training batch size *miniBatchFraction*, and regularization coefficient *regParam*. To sum up, the best combination of parameters is iterations = 100, step = 10, miniBatchFraction = 0.1, regParam0.001, regType = l2, and

Table 2 SVM model evaluation

```
# define the model evaluation function
def ModelAccuracy(model, validationData):
   # The accuracy of the calculation model
   predict = model.predict(validationData.map(lambda
   p:p.features))
   predict = predict.map(lambda p: float(p))
   # splicing predicted and actual values
   predict_real= predict.zip(validationData.map(lambda p:
   p.label))
   matched = predict_real.filter(lambda p:p[0]==p[1])
   accuracy = float(matched.count()) /
   float(predict_real.count())
   return accuracy

acc = ModelAccuracy(logModel, validationData)
# print accuracy
print("accuracy="+str(acc))
```

the corresponding accuracy. We applied this selection of values when implementing our model in a predictive situation. The trace of maintenance activities shows that our learning phase targeted the preventive restart of business service at quasi-fixed period. We have raised maintenance tickets for these services whose instability seems obvious. An extraction of the associated log data highlights memory limit concerns and for other resource access conflicts and abnormal waiting times. These interventions lead to a decrease in the number of application failures, and we can now evaluate the uptime of the applications studied.

6 Conclusion

This study provides several useful insights for our continued Big Data computing cases. First, we have shown that another Big Data orchestration architecture is possible instead of the traditional Cloudera-like cluster based on preconfigured virtual machines. We defined a Kubernetes orchestration based on service images useful for our case study and useful only by itself. Finally, we exploited this architecture where each component is isolated from its neighbors but shares the same network in order to communicate between them and obtain the construction of an AI model for the definition of maintenance operations on our application servers. We obtained a clear result of reducing incidents based on restarting identified services. A greater availability of business applications is highlighted and leads us to continue this work after a redesign of services frequently causing incidents.

References

1. Bilal, M., Oyedele, L.O., Akinade, O.O., Ajayi, S.O., Alaka, H.A., Owolabi, H.A., et al.: Big data architecture for construction waste analytics (CWA): a conceptual framework. J. Build. Eng. **6**, 144–156 (2016)
2. Matsebula, F., & Mnkandla, E. (2017, September). A big data architecture for learning analytics in higher education. In 2017 IEEE AFRICON, pp. 951–956. IEEE
3. Jones, K.M.L.: Learning analytics and higher education: a proposed model for establishing informed consent mechanisms to promote student privacy and autonomy. Int. J. Educ. Technol. High. Educ. **16**, 24 (2019). https://doi.org/10.1186/s41239-019-0155-0
4. Wang, J., Yang, Y., Wang, T., Sherratt, R.S., Zhang, J.: Big data service architecture: a survey. J. Int. Technol. **21**(2), 393–405 (2020)
5. Huber, M.F., Voigt, M., Ngomo, A.C.N.: Big data architecture for the semantic analysis of complex events in manufacturing. In: Informatik 2016, p. 353. Gesellschaft für Informatik e.V, Bonn (2016)
6. Santos, M.Y., e Sá, J.O., Costa, C., Galvão, J., Andrade, C., Martinho, B., et al.: A big data analytics architecture for industry 4.0. In: World conference on information systems and technologies, pp. 175–184. Springer, Champions (2017, April)
7. Ivanov, T., Singhal, R.: Abench: Big data architecture stack benchmark. In: Companion of the 2018 ACM/SPEC International Conference on Performance Engineering, pp. 13–16 (2018, April)

8. Sassi, M.S.H., Jedidi, F.G., Fourati, L.C.: A new architecture for cognitive internet of things and big data. Procedia Comput. Sci. **159**, 534–543 (2019)
9. Mokhtari, G., Anvari-Moghaddam, A., Zhang, Q.: A new-layered architecture for future big data-driven smart homes. IEEE Access. **7**, 19002–19012 (2019)
10. Jindal, R., Taneja, S.: U-struct: a framework for conversion of unstructured text documents into structured form. In: International Conference on Advances in Computing, Communication and Control, pp. 59–69. Springer, Berlin/Heidelberg (2013, January)

Intelligent Target Cell Selection Algorithm for Low Latency 5G Networks

Vincent Omollo Nyangaresi, Mustafa A. Al Sibahee, Zaid Ameen Abduljabbar, Abdulhadi Alhassani, Iman Qays Abduljaleel, and Enas Wahab Abood

1 Introduction

At cell boundaries, the user equipment (UE) may connect back and forth between multiple cells [1] due to reception of signals from several cells. As such, quality of service (QoS) may be reduced in these regions since these ping-pong handovers

V. O. Nyangaresi (✉)
Faculty of Biological & Physical Sciences, Tom Mboya University College, Homabay, Kenya
e-mail: vnyangaresi@tmuc.ac.ke

M. A. Al Sibahee
College of Big Data and Internet, Shenzhen Technology University, Shenzhen, China

Computer Technology Engineering Department, Iraq University College, Basrah, Iraq
e-mail: mustafa@sztu.edu.cn

Z. A. Abduljabbar
Department of Computer Science, College of Education for Pure Sciences, University of Basrah, Basrah, Iraq

Huazhong University of Science and Technology, Shenzhen Institute, Shenzhen, China
e-mail: zaid.ameen@uobasrah.edu.iq

A. Alhassani
Communications Engineering Department, Iraq University College, Basrah, Iraq
e-mail: hadi.alhassani@iuc.edu.iq

I. Q. Abduljaleel
Department of Computer Science, College of Computer Science and Information Technology, University of Basrah, Basrah, Iraq
e-mail: iman.abduljaleel@uobasrah.edu.iq

E. W. Abood
Department of Mathematics, College of Science, University of Basrah, Basrah, Iraq
e-mail: enas.abood@uobasrah.edu.iq

M. D. Hina et al. (eds.), *Advances in Computational Intelligence and Communication*, EAI/Springer Innovations in Communication and Computing,
https://doi.org/10.1007/978-3-031-19523-5_6

cause frequent call drops and high latencies. Consequently, target cell selection is very critical during cellular network handover process, more so in 5G networks where long latencies are undesirable [2]. In particular, there is need for the incorporation of ideal parameters during the handover decision process so as to boost both QoS and quality of experience (QoE) after the handover process. In legacy cellular systems, received signal strength indicator (RSSI) is a very popular handover parameter. However, using only RSSI serves to increase the probability of handoff failure, handover latency, and packet losses [3]. In [4], the authors point out that this may also result in ping-pong handovers, potentially causing longer delays in cellular networks. This problem can be addressed by the incorporation of additional parameters during the handover decision process.

As explained in [5], most of the conventional handover techniques are reactive in nature. This means that control signaling and channel assignment are only executed when the handoff process is initiated. Obviously, this leads to elongated delays, which contradict the ultra-reliability and extremely low delays 5G requirements. To address these shortcomings, a need arises for the development of proactive handover strategies. In these approaches, control signaling and channel assignment or reservation is executed long before the actual handover initiation phase. To achieve this, the user equipment's next point of attachment needs to be precisely predicted in advance [6].

Security and privacy protection are other key challenging issues during the handover process in 5G networks. In [7], it is explained that lack of authentication during handovers exposes the cellular network to numerous attacks. Some of these attacks and threats include impersonation man-in-the-middle (MitM), eavesdropping, and de-synchronization. In a typical cellular network deployment scenario, 5G co-exists with other cellular technologies such as 4G, 3G, or even 2G. As such, it is possible for the UE to be handed over from 5G networks to these other networks, which are less secure. For instance, although 3G implements handover authentication, the deployed protocol lacks backward-compatibility support and is prone to redirection and MitM attacks [8].

According to [9], 5G networks deploy very high frequencies, which result in extremely poor signal propagations such as high signal attenuations. This is may be due to signal obstructions by physical obstacles such as buildings, human bodies, and reflectors. In the final analysis, there is channel intermittency, which can lead to high packet drop rates. The rapid mobility of the user equipment among the small cells in 5G networks implies frequent handoffs between various attachment points. For improved security, these frequent handoffs need to be properly authenticated. As such, there is increased latency during packet delivery, which goes against the goals of 5G [10–11]. The support of a myriad of services and devices implies that 5G networks have to support interoperability among these services and devices. Since these devices are required to authenticate each other, the 5G networks have higher security and privacy requirements compared with their predecessors such as 3G and 2G [12].

In 5G networks, authentication is via improved EAP-AKA protocol, which has several security issues such as simulated attacks, MitM attacks, and lack of user identity privacy protection [13]. In addition, this protocol has high bandwidth consumption, heavy signaling overheads, and slow re-authentication during handover. According to [14], the deployment of inefficient protocols for authenticating 5G handovers always degrades performance and increases overall network latencies. In addition, there is a high handoff failure rate in these networks, which serves to further worsen the QoS [15].

As discussed in [11], the provision of adequate privacy for user data and effective resource management are other challenges that need to be addressed in 5G networks. Authors in [16] stress on the significance of providing secure and private communications, as well as low delays during the handoff process. Typically, strong authentication is the first step toward security in communication networks [17]. To ensure enhanced cellular networks security, strong authentication should be robust against conventional cellular network attacks such as privileged insider, de-synchronization, password guessing, and impersonations [18]. In addition, the authentication scheme should uphold local verification and user anonymity. According to [19], anonymous communication is another important privacy goal that should be upheld during the handoff procedures. In this article, the contributions below are acclaimed:

- An algorithm that deploys ANN-FL is developed for efficient target cell selection within the hysteresis region.
- A novel handover authentication scheme is developed based on lightweight hashing and symmetric encryptions.
- Formal security analysis is executed to demonstrate that the proposed approach offers sufficient and strong mutual authentication among the user equipment, source gNB, and target gNB.
- Informal security analysis is carried out to show that this algorithm provides salient security features such as backward and forward secrecy. In addition, it is demonstrated to be robust against a number of conventional 5G network attacks.

The rest of this paper is structured as follows: Sect. 2 presents a discussion of the related literature, while Sect. 3 gives an illustration of the proposed algorithm. Conversely, Sect. 4 discusses the obtained results, while Sect. 5 provides the conclusion of this paper.

2 Related Work

Target cell selection and security during handovers in cellular networks have attracted much attention from academia and industry. For instance, a fuzzy logic-based handoff scheme is presented in [20]. On the other hand, an energy-efficient handoff protocol is developed in [21]. In this scheme, three input variables including UE speed, battery level, and QoE are incorporated in the handover decision process. Similarly, machine learning-based handoff protocols are presented in [22, 23, 25].

An updated ant colony optimization protocol is presented in [26] for the facilitation of seamless handoffs. To determine the best target network, all the available networks are buffered in an archive table, and the cost function is computed using parameters of all the networks such as monetary cost, power consumption, RSS, bandwidth, and security. In addition, a handover algorithm to facilitate target selection and ensure load balancing among network devices is presented in [8]. On the other hand, RSS-based neural network-based scheme is developed in [27] for the prediction of number of subscribers in the network.

In terms of security, a multi-server-based authentication protocol is developed in [28], while a ticket-based handoff scheme is presented in [29]. However, the scheme in [29] is insecure due to transmission of handover parameters in plaintext. As such, improved group key security techniques are developed in [17, 30]. Unfortunately, the presence of malicious access points within the group may compromise the entire system [31]. A smart card-based user authentication protocol is introduced in [32], while a cloud-based 5G authentication scheme is developed in [33]. However, the protocol in [33] cannot offer perfect forward secrecy and is also susceptible to privileged insider attacks. On its part, the protocol in [34] does not guarantee user anonymity and is subject to offline password-guessing attacks. A novel anonymous roaming authentication scheme is presented in [18], while a certificate-based scheme is introduced in [9] for 5G handoff authentication. However, certificate management in [9] is a bit cumbersome. Similarly, a certificate authority (CA)-based lightweight intelligent authentication protocol is presented in [35] to counter de-synchronization attack, man-in-the-middle attacks, and minimize handoff latencies. In addition, CA-based scheme has been introduced in [36] to address vulnerabilities and provide handover key management.

3 Proposed Algorithm

The proposed algorithm comprised of two major functions, target selection and handover authentication. Here, target selection helped identity the most probable cell to handover the UE to, based on measured figures of merit (FOMs). On the other hand, strong mutual authentication and key negotiation are required to uphold the security and privacy of the exchanged packets during the handover process.

3.1 Target Cell Selection

To facilitate efficient target cell selection, a combination of artificial neural network and fuzzy logic is deployed. ANN dynamically adjusted the handover conditions to facilitate handover instant determination. On the other hand, FL helped determine the best target cell for handover. As shown in Algorithm 1, the first step was to partition the coverage area into three concentric regions. These tracking regions included no handover region (NHR), low probability handover region (LPHR), and high probability handover region (HPHR). The rationale for this partitioning as well as the technical details of ANN-FL is well articulated in [37]. Afterward, the figures of merit, which comprised of blocking probability (P_b), traffic intensity (A_C), power density (P_D), received carrier power (P_r), and path loss (P_L), were measured. As shown in Algorithm 1, the user equipment within the NHR has sufficiently better QoS, and hence, neighbor scanning is not executed in this region. However, its mobility parameters need to be traced (steps 2 and 4). On the other hand, the QoS at the LPHR are relatively low, and hence, neighbor cell beacons need to be scanned and buffered in the source gNB (SgNB) handover decision matrix (HDM) database (steps 7 and 8). For the UE located at the HPHR, the time to trigger (TTT) parameter is satisfied, and the proposed ANN-FL system is invoked to decide the necessity of the handover as well as the selection of the most ideal target gNB (steps 11 to 19). Once in the target gNB (TgNB), the adaptive neural network is utilized to adaptively adjust the handoff hysteresis and hence modify the output of the fuzzy logic inference accordingly (step 20). Here, if handover factor $_i$ is greater than handover factor $_S$, then event A2 is re-triggered followed by event A3, and the UE is handed over to the best available TgNB (step 19); otherwise step 22 is invoked to evaluate the next rule. In step 23, event A1 is triggered whenever $_i$ is less than $_S$.

Algorithm 1: Target Cell Selection

BEGIN:

1. Segment coverage area, A_{hex} into three concentric regions
2. **FOR** UE within NHR **DO:**
3. Track UE movement coordinates
4. Never scan neighbouring gNBs figures of merit
5. **ENDFOR**
6. **FOR** UE within LPHR **DO:** /* *Event A2 triggered to start measurements* */
7. Measure P_b, A_C, P_r , P_D, P_L in current and neighboring gNBs
8. Buffer the FOM in (7) in the SgNB HDM
9. **ENDFOR**
10. **FOR** UE within HPHR **DO:**
11. Initialize number of rules, N in the knowledge base
12. Compute handover threshold factor of serving gNB, $_S$
13. **FOR** $i < N$ **DO:**
14. Fuzzify the FOM in the SgNB and pass them to inference engine
15. Apply the rules in the knowledge base to the FOM
16. De-fuzzify the results in (15) above and select the best target gNB
17. Calculate handover factor, $_i$ of the current rule i in the knowledge base
18. **IF** $_i > _S$ **THEN** /* *hysteresis margin exceeded, Re-triggering event A2, then event A3* */

19. Handover the UE to the target gNB in (16)
20. Monitor FOM in the new cell and modify output of (15) accordingly
21. **ELSE**
22. Evaluate the next rule in the knowledge base
23. **IF** $i = N$ and $_i < _S$ **THEN:** /* *hysteresis margin exceeded , event A1 triggered**/
24. Stay in the source gNB
25. **ENDIF**
26. **ENDIF**
27. **ENDFOR**
28. **ENDFOR**

END

3.2 Handover Authentication and Key Agreement

The UE, SgNB, and TgNB are the handover entities that are actively involved in this protocol. However, the anchor mobility function (AMF) is involved toward the end of the key agreement phase for handling path switch requests. Here, SgNB is the base station currently in charge of the UE's radio communication. On the other hand, TgNB denote the base station whose FOMs have been determined to be superior to those of the current cell. The deployed notations are presented in Table 1 below.

The first step in the proposed algorithm is the setting of the threshold timestamp $\Delta\Upsilon$ for reply attack prevention as shown in Algorithm 2. In step 2, the UE construct beacon $_1$ consisting of nonce ŋ, timestamp Υ, and authentication request $Auth_{Req}$, protected by session keys employed between the UE and SgNB ( ) and session key used between the TgNB and UE (ђ).

Algorithm 2: Handover Authentication and Key Agreement

BEGIN:

*/*Begin Authentication*/*

1. Set $\Delta\Upsilon$
2. Compose $_1$: $Auth_{Req}$ {(ŋ,Υ)ђ, Υ}
3. UE→ SgNB: { $_1$}
4. **IF** $\Upsilon > \Delta\Upsilon$ **THEN**: Flag as replay & terminate
5. **ELSE:**

   Compose $_2$: {(ŋ$_1$,Υ)ђ, Υ_1, }

6. SgNB→ TgNB: { $_2$}
7. Using , decrypt $_2$
8. **IF** $\Upsilon_1 > \Delta\Upsilon$ **THEN**: Flag as replay & terminate

9. **ELSE:**

Extract NH from

10. Compute $\hbar^{*}$ = KDF (NH||PCI||ARFCN-DL)
11. Using $\hbar^{*}$, decrypt $\{(\text{ŋ},\Upsilon)_{\hbar}\}$
12. **IF** $\Upsilon > \Delta\Upsilon$ **THEN**: Flag as replay& terminate
13. **ELSE:**
14. Compose H_{ACK}: $\{\text{ŋ}_2, \Upsilon_2\}_{\hbar}$
15. TgNB→ UE: $\{H_{ACK}\}$
16. **IF** $\Upsilon_2 > \Delta\Upsilon$ **THEN**: Flag as replay & terminate
17. **ELSE:**
18. **IF** $\text{ŋ}_2 != \text{ŋ}$ **THEN**: Flag as malicious & terminate
19. **ELSE:**
20. Trust both SgNB & TgNB */*End of Authentication*/*
21. **ENDIF**
22. **ENDIF**
23. **ENDIF**
24. **ENDIF**
25. **ENDIF**
26. TgNB→ AMF: $\{PS_{Req}\}$
27. Compute *, ß* & compose PS_{ACK}: { *, Υ_3, {ß*, Υ_3}K_{AMF}}
28. AMF→ TgNB: $\{PS_{ACK}\}$
29. Buffer { *}K_{AMF} & compose $_{3}$: {{ß*, Υ_3}K_{AMF}, Υ_4}$_{\hbar}$
30. TgNB →UE: { $_{3}$}

END

In step 3, beacon { $_{1}$} is sent to SgNB where it is decoded to extract Υ for verifying the freshness of beacon $_{1}$. If it is more than $\Delta\Upsilon$, the beacon is flagged as replay (step 4); otherwise, $_{2}$ is constructed (step 5) before being sent to the TgNB (step 6). Here, $_{2}$ contains that masks NH, and all the parameters are protected by the session key used between SgNB and TgNB, . Upon receiving the handover request, the TgNB decrypts $_{2}$ using (step 7) before checking the freshness of $_{2}$ using new timestamp Υ_1 to thwart any replays. Provided that $_{2}$ is fresh, TgNB extracts NH from (step 9). Thereafter, the obtained NH is utilized to derive $\hbar^{*}$ (step 10) before being used to decrypt beacon $\{(\text{ŋ},\Upsilon)_{\hbar}\}$ as shown in step 11. Afterward, the freshness of the decrypted message is verified using the obtained timestamp Υ (step 12) such that if it is valid, handover acknowledgment H_{ACK} is generated, consisting of nonce ŋ_2 and new timestamp Υ_2 (step 14). In step 15, this acknowledgment is transmitted to the user equipment. On receiving H_{ACK}, the user equipment verifies its freshness (step 16) followed by validation of nonce ŋ_2 against nonce ŋ (step 18). Provided that both freshness and nonce are valid, all the three entities would have authenticated one another and hence can trust each other. To prepare for the next handover, PS_{Req} is transmitted over to the AMF (step 26). Next, new * and ß* are computed in accordance with step 27. Afterward, using K_{AMF} and , the AMF composes PS_{ACK} consisting of encrypted *, ß*, and timestamps

Table 1 Notations and descriptions

Notation	Description
KDF	Key derivation function
NH	Next hop
PCI	Physical cell identity
ARFCN-DL	E-UTRAN absolute radio Frequency Channel number on the downlink
NCC	Next chaining counter
K_{AMF}	AMF's master key
SgNB	Source gNB
TgNB	Target gNB
H_{ACK}	Handover acknowledgment
PS_{Req}	Path switch request
PS_{ACK}	Path switch acknowledgement
ɣ	UE cipher key
ŋ, $ŋ_1$, $ŋ_2$	Random nonces
, ß	Parameters to secretly send NH and NCC
$\Delta\Upsilon$	Threshold timestamp
$\Upsilon, \Upsilon_1, \Upsilon_{2,} \Upsilon_3, \Upsilon_4$	Timestamps
Þ	AMF-TgNB session key
	UE-SgNB session key
ђ	UE-TgNB session key
	SgNB-TgNB session key
	SgNB-AMF session key
	TgNB-AMF session key
||	Concatenation operator

such that only the UE can decipher it. This PS_{ACK} is then sent over to the TgNB in step 28. Here, * is stored followed by the construction of 3 (step 29). To finalize this phase, $_3$ is forwarded to the UE in step 30.

4 Results and Discussion

In this section, the results are presented based on the offered security features as well as the performance of the proposed scheme.

4.1 Security Evaluation

This subsection presents both the formal and informal security analysis of the proposed scheme. Here, formal security analysis is executed through the Burrows-Abadi-Needham (BAN) logic, which is the most popular technique for evaluating authentication protocols. Thereafter, informal analysis is carried out to show that the proposed scheme is resilient against typical attacks in the 5G networks.

4.1.1 Formal Security Analysis

To accomplish this, the notations in Table 2, BAN logic rules in Table 3, and assumptions in Table 4 are utilized. As shown in Table 2, nine BAN logic notations were critical for the formal analysis of the proposed algorithm.

A shown in Table 2, nine notations are crucial during this formal security analysis, while Table 3 shows that seven rules are critical for this formal verification.

To facilitate easier analysis of the proposed algorithm, the 14 initial assumptions in Table 4 were formulated. These assumptions were based on Algorithm 2 and are derived in accordance with the shared secret keys as well as the exchanged messages.

To proof that the proposed algorithm provides salient security features, four security goals that are significant during the authentication and key agreement process are stated in Table 5.

Table 2 BAN logic notations

Notation	Description
$X\|\equiv Y$	X believes Y
$X \lhd Y$	X sees Y
$X\| \sim Y$	X said Y
$X \Rightarrow Y$	X has complete jurisdiction over Y
$\#(Y)$	Y is fresh
$\langle Y \rangle_F$	Y is combined with F
$\{Y\}_F$	Y is encrypted under key F
$X \overset{L}{\leftrightarrow} G$	X and G share key L
$A \overset{L}{\leftrightharpoons} B$	L is a secret known by A and B

Table 3 BAN logic rules

Rule	Description
$\frac{A\|\equiv\#(C)}{A\|\equiv\#(C,D)}$	Fresh promotion rule (FPR)
$UE \|\equiv UE \overset{\mathfrak{h}}{\leftrightarrow} TgNB$	Message meaning rule (MMR)
$\frac{A\|\equiv A \overset{D}{\leftrightharpoons} B, A\langle C \rangle_D}{A\|\equiv B\|\sim c}$	Message meaning rule with a shared secret (MMR - SS)
$\frac{A\|\equiv\#(C), A\|\equiv B\|\sim C}{A\|\equiv B\|\equiv C}$	Nonce-verification rule (NVR)
$\frac{A\|\equiv B \Rightarrow C, A\|\equiv B\|\equiv C}{A\|\equiv C}$	Jurisdiction rule (JR)
$TgNB \|\equiv TgNB \overset{\mathfrak{h}}{\leftrightarrow} UE$	Decomposition rule (DR)
$\frac{A\|\equiv C, A\|\equiv D}{A\|\equiv(C,D)}$	Composition rule (CR)

Table 4 BAN logic initial assumptions

SNo.	Description
S1	$UE \mid\equiv UE^{TgNB \mid\equiv UE \mid\equiv UE \overset{\text{ђ}}{\leftrightarrow} TgNB}_{SgNB}$
S2	$UE \mid\equiv UE \overset{\gamma}{\leftrightharpoons} AMF$
S3	$UE \mid\equiv AMF \Rightarrow NCC$
S4	$SgNB \mid\equiv SgNB^{UE \mid\equiv TgNB \mid\equiv TgNB \overset{\text{ђ}}{\leftrightarrow} UE}_{TgNB}$
S5	$SgNB \mid\equiv SgNB \overset{\phi}{\leftrightarrow} AMF$
S6	$SgNB \mid\equiv AMF \Rightarrow$
S7	$SgNB \mid\equiv AMF \Rightarrow \text{ß}$
S8	$SgNB \mid\equiv SgNB \frac{A \mid\equiv \#(C)}{A \mid\equiv \#(C,D)} UE$
S9	$TgNB \mid\equiv TgNB \frac{A \triangleleft (C,D)}{A \triangleleft C}, \frac{A \mid\equiv A \overset{k}{\leftrightarrow} B, A \triangleleft \{C\}_k}{A \triangleleft C} SgNB$
S10	$TgNB \mid\equiv TgNB \overset{\lambda}{\leftrightarrow} AMF$
S11	$TgNB \mid\equiv TgNB \overset{\psi}{\leftrightharpoons} AMF$
S12	$TgNB \mid\equiv AMF \Rightarrow \text{ђ}$
S13	$UE \mid\equiv UE \overset{K_{AMF}}{\leftrightarrow} AMF$
S14	$UE \mid\equiv AMF \Rightarrow \text{ß}$

Table 5 Proposed algorithm security goals

SNo.	Security goal
G-1	$UE \mid\equiv UE^{UE \mid\equiv UE \overset{\beta}{\leftrightarrow} SgNB}_{TgNB}$
G-2	$TgNB \mid\equiv TgNB^{SgNB \mid\equiv SgNB \overset{\eth}{\leftrightarrow} TgNB}_{UE}$
G-3	$UE \mid\equiv TgNB \mid^{SgNB \mid\equiv SgNB \overset{\beta}{\leftrightarrow} UE}_{UE}$
G-4	$TgNB \mid\equiv UE \mid\equiv UE^{TgNB \mid\equiv TgNB \overset{\eth}{\leftrightarrow} SgNB}_{TgNB}$

Table 6 Exchanged messages

Message	Entities involved
Msg1	$UE \rightarrow SgNB$: $\{\ _1\}$
Msg2	$SgNB \rightarrow TgNB$: $\{\ _2\}$
Msg 3	$TgNB \rightarrow UE$: $\{H_{ACK}\}$
Msg 4	$AMF \rightarrow TgNB$: $\{PS_{ACK}\}$
Msg 5	$TgNB \rightarrow UE$: $\{\ _3\}$

Based on Algorithm 2, the communicating entries exchange five messages during the authentication and key negotiation procedures. These communicating entities comprise user equipment, source gNB, target gNB, and the achor mobility function. The five exchanged messages are shown in Table 6.

Since the last two messages are signaling beacons for the subsequent handover, the assumption made here is that they had already been exchanged before handover initiation. Therefore, the current SgNB was the TgNB during the previous handover.

Table 7 Idealized exchanged messages

Message	Idealized form
Msg1	AMF → SgNB:{ ,Υ_3^{Prev},{ß,Υ_3^{Prev}}K_{AMF}}
Msg2	SgNB →UE: {{ß, Υ_3^{Prev}}K_{AMF}, Υ_4^{Prev}}$_{ђ}$
Msg 3	UE → SgNB: {ϒ, {ŋ, ϒ, ђ}$_{ђ}$}
Msg 4	SgNB→ TgNB: {$ϒ_1$, ,{$ŋ_1$,ϒ, ђ}$_{ђ}$}
Msg 5	TgNB→ UE: {$ŋ_2$, $ϒ_2$, ђ}$_{ђ}$

Denoting previous handover elements as $(.)^{Prev}$, the message exchanges are normalized as shown in Table 7.

The formal analysis using BAN logic proofs (BLPs) then proceeds as follows:

Based on Msg1:

BLP_1: SgNB ◁{ ,Υ_3^{Prev},{ß,Υ_3^{Prev}}K_{AMF}} .

According to S5, MMR is applied to yield BLP_2:

BLP_2: SgNB |≡AMF|~{ ,Υ_3^{Prev},{ß,Υ_3^{Prev}}K_{AMF}}.

The SgNB checks the freshness of Υ_3^{Prev} and if it is:

BLP_3: SgNB |≡#(Υ_3^{Prev}).

Based on BLP_3, FPR is employed to obtain BLP_4:

BLP_4: SgNB |≡#(,Υ_3^{Prev},{ß,Υ_3^{Prev}}K_{AMF}).

According to BLP_2 and BLP_4, NVR is applied to get BLP_5:

BLP_5: SgNB |≡AMF|≡{ , {ß,Υ_3^{Prev}}K_{AMF}}.

Based on S6, S7 and BLP_5, the application of JR yields BLP_6:

BLP_6: SgNB |≡{ , {ß,Υ_3^{Prev}}K_{AMF}}.

According to Msg 2:

BLP_7: UE ◁{{ß,Υ_3^{Prev}}K_{AMF},Υ_4^{Prev}}.

Based on BLP_7 and S13, application of MMR results in BLP_8:

BLP_8: UE |≡AMF|~{ß,Υ_3^{Prev}}.

Afterward, the UE validates message freshness:

BLP_9: UE |≡ #(Υ_4^{Prev}).

Applying FPR to BLP_8 yields BLP_{10}:

BLP_{10}: UE |≡ #({ß, Υ_3^{Prev}}).

On the other hand, using NVR on both BLP_8 and BLP_{10} results in BLP_{11}:

BLP_{11}: UE |≡AMF|≡{ß,Υ_3^{Prev}}.

Using DR on BLP_{11} yields BLP_{12}:

BLP_{12}: UE |≡AMF|≡ß

On the other hand, based on S14, JR is used on BLP_{12} to get BLP_{13}:

BLP_{13}: UE |≡ß

Because the UE is capable of extracting NCC from ß and utilize it to derive ђ, then:

BLP_{14}: UE |≡ ђ

Consequently, UE |≡ UE$^{UE \mid\equiv UE \overset{ђ}{\leftrightarrow} TgNB}$TgNB, and hence G-1 is achieved.
Based on Msg 3:

BLP_{15}: SgNB ◁{ϒ, {ŋ, ϒ, ђ}$_{ђ}$} .

Based on S8, MMR is applied on BLP_{15} to obtain BLP_{16}:

BLP_{16}: SgNB |≡ UE|~{ϒ, {ŋ, ϒ, ђ}$_{ђ}$}.

Thereafter, SgNB validates the freshness of ϒ such that if it is fresh, then:

BLP_{17}: SgNB |≡#(ϒ).

The application of FPR to BLP_{17} results in BLP_{18}:

BLP_{18}: SgNB |≡{ŋ, ϒ, ђ}$_{ђ}$.

Using NVR on both BLP16 and BLP18 yields BLP_{19}:

BLP_{19}: SgNB |≡UE| ≡ {ŋ, ϒ, ђ}$_{ђ}$.

Based on Msg 4:

BLP_{20}: TgNB ◁{ϒ$_1$, , {ŋ$_1$,ϒ, ђ}$_{ђ}$} .

Based on S9, MMR is applied in BLP_{20} to obtain BLP_{21}:

BLP_{21}: TgNB |≡SgNB|~{ϒ$_1$, , {ŋ$_1$,ϒ, ђ}$_{ђ}$}.

Thereafter, TgNB verifies the freshness of ϒ$_1$ such that it is fresh, then:

BLP_{22}: TgNB |≡#(ϒ$_1$).

Applying FPR in BLP_{22} results in BLP_{23}:

BLP_{23}: TgNB |≡#(, {ŋ$_1$,ϒ, ђ}$_{ђ}$).

On the other hand, using DR in BLP_{23} yields BLP_{24}:

BLP_{24}: TgNB |≡#( ).

Considering the fact that is derived with Ψ, and TgNB can derive ђ using the data extracted from , then:

BLP_{25}: $TgNB \lhd \langle\mathfrak{h}\rangle_{\Psi}$.

Based on S11, MMR-SS is applied to BLP_{25} to get BLP_{26}:

BLP_{26}: $TgNB \mid\equiv AMF \mid\sim \mathfrak{h}$

Applying NVR to both BLP_{24} and BLP_{26} results in BLP_{27}:

BLP_{27}: $TgNB \mid\equiv AMF \mid\equiv \mathfrak{h}$

According to S12, JR is used in BLP_{27} to yield BLP_{28}:

BLP_{28}: $TgNB \mid\equiv \mathfrak{h}$

As such, $TgNB \mid\equiv TgNB^{TgNB \mid\equiv TgNB \overset{\mathfrak{h}}{\leftrightarrow} UE} UE$, and hence G-2 is realized.

According to S9, DR is used in BLP_{20} to get BLP_{29}:

BLP_{29}: $TgNB \lhd \{\eta_1, \Upsilon, \mathfrak{h}\}_{\mathfrak{h}}$

On the other hand, MMR is used in both BLP_{28} and BLP_{29} to get BLP_{30}:

BLP_{30}:$TgNB \mid\equiv UE \mid\sim \{\eta_1, \Upsilon, \mathfrak{h}\}$

Thereafter, TgNB verifies message freshness using Υ such that if it is fresh, then:

BLP_{31}: $TgNB \mid\equiv \#(\Upsilon)$

The application of FPR in BLP_{31} yields BLP_{32}:

BLP_{32}: $TgNB \mid\equiv \#(\mathfrak{h})$

Using NVR on both BLP_{30} and BLP_{32} yields BLP_{33}:

BLP_{33}: $TgNB \mid\equiv UE \mid\equiv \mathfrak{h}$

Consequently, $TgNB \mid\equiv UE \mid\equiv UE^{UE \mid\equiv TgNB \mid\equiv TgNB \overset{\mathfrak{h}}{\leftrightarrow} UE} TgNB$, and hence G-4 is attained.

According to Msg 5:

BLP_{34}: $UE \lhd \{\eta_2, \Upsilon_2, \mathfrak{h}\}_{\mathfrak{h}}$

The application of MMR to both BLP_{14} and BLP_{34} results in BLP_{35}:

BLP_{35}: $UE \mid\equiv TgNB \mid\sim \{\eta_2, \Upsilon_2, \mathfrak{h}\}$

Thereafter, the UE validates the freshness of Υ_2 such that if it is fresh:

BLP_{36}: $UE \mid\equiv \#(\eta_2, \Upsilon_2, \mathfrak{h})$

This is followed by UE's verification of the nonce in Msg 5 such that if it is dissimilar to that in Msg 3, the message is marked as malevolent and is therefore discarded. On the other hand, if the two nonces are identical, then the following steps are invoked:

Applying NVR to both BLP_{35} and BLP_{36} yields BLP_{37}:

BLP_{37}: $UE \mid\equiv TgNB \mid\equiv \{\eta_2, \Upsilon_2, \mathfrak{h}\}$

Finally, DR is used in BLP_{37} to result in BLP_{38}:

BLP_{38}: UE |≡TgNB|≡ђ

As such, UE |≡TgNB|TgNB |≡ UE |≡ $UE \overset{ђ}{\leftrightarrow} TgNB$UE, thus G-3 is attained.

Clearly, BLPs 1 to BLPs 38 have demonstrated that this algorithm attains all the four security goals that were formulated. As such, the AMF has facilitated strong perfect authentication and key negotiation among the UE, SgNB, and TgNB.

4.1.2 Informal Security Analysis

A number of attacks exist in the 5G network authentication protocols as well as in other protocols that have been developed recently to curb weaknesses in 5G networks. The following security features and attack models were found to be critical in 5G networks, and hence, the proposed scheme is evaluated as discussed below.

Provision of Forward and Backward Key Secrecy The proposed algorithm derived ђ through one-way hash function using NH, PCI, and ARFCN-DL as inputs. Here, TgNB has no knowledge of and can never compute it. In addition, the SgNB has no knowledge of ђ and can never derive it based on the messages it receives from TgNB. As such, it is infeasible for an adversary to successfully derive both ђ and from each other. As such, both forward and backward security are assured.

Resilience Against Packet Replays The aim of the adversary is to compromise a legitimate gNB so that the UE can handover to it and subsequently enable the modification and forwarding of bogus messages to gNBs. Although the attacker may receive from the AMF, any attempt to change it implies that TgNB will never get the correct NH to facilitate the computation of ђ. Consequently, TgNB is unable to obtain the correct value of ŋ and ϒ. In addition, the decrypted ϒ will fail the freshness check, while the value of ŋ will be dissimilar to $ŋ_2$, leading to flagging of the message as malicious and subsequent dropping of the authentication request. This implies that the UE will never receive H_{ACK} from TgNB and will know that the handover has failed. As such, replay attack is easily detected in the proposed algorithm.

Robustness Against De-synchronization Attacks The goal of the attacker here is to utilize invalid NCC value to misinform the UE. In the proposed algorithm, NCC is securely masked in ß whose security is assured through K_{AMF}. Since K_{AMF} is only known to the UE and AMF, neither the TgNB nor SgNB can obtain ß.

The resilience of the proposed algorithm to both de-synchronization and replay attacks renders it superior to the conventional 5G AKA handover protocol.

Table 8 Handover and packet losses

Simulation Iterations	Conventional RSSI protocol		Proposed algorithm	
	Latency (seconds)	Packet losses (bytes)	Latency (seconds)	Packet losses (bytes)
1	0.070	49	0.030	30
2	0.055	45	0.043	40
3	0.110	54	0.028	28
4	0.072	50	0.049	43
5	0.067	49	0.019	13
6	0.092	53	0.039	38
7	0.065	48	0.024	23
8	0.045	41	0.037	36
9	0.089	52	0.027	26
10	0.060	47	0.022	18

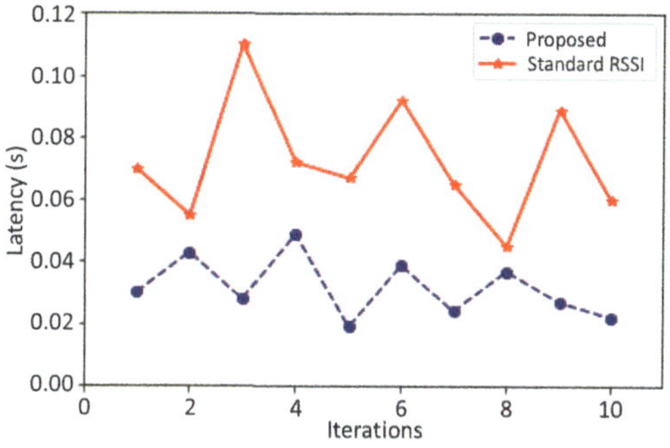

Fig. 1 Latencies for RSSI and proposed algorithm

4.2 *Performance Evaluation*

In typical handover protocols, handover latencies, packet losses, and computational complexities are the most frequently deployed metrics during performance evaluations. The obtained metrics using the developed scheme are therefore compared with that of related schemes, including the legacy RSSI-based handover decision algorithm.

Latencies and Packet Losses To execute these evaluations, both RSSI and the proposed algorithm were iterated ten times as shown in Table 8. This number of iterations was experimentally determined to be sufficient for the measurement of both RSSI and proposed algorithm's latencies and packet losses.

As shown in Table 8, the handover delays and packet loses varied for different simulation iterations in both RSSI protocol and the proposed scheme. Based on the results in Fig. 1, standard RSSI protocol had the longest latency of 0.11 s and the shortest latency of 0.045 s. However, the longest latency for the proposed scheme is 0.049 s, while the shortest latency is 0.019 s.

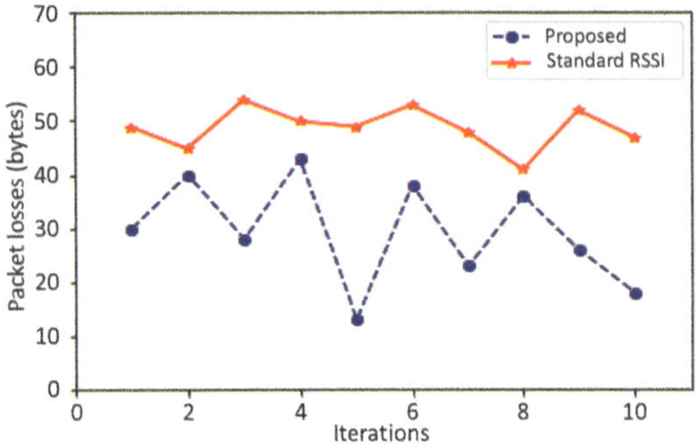

Fig. 2 Packet losses for RSSI and proposed algorithm

Table 9 Computation cost comparisons

Scheme	Computation costs (ms)
[28]	0.20696
[33]	0.18626
[34]	0.11899
Proposed	0.019922

On average, the standard RSSI protocol requires 0.0725 s for complete execution, while the proposed algorithm requires only 0.0318 s. This represents a 56.1% reduction in handover delays. Diagrammatically, the relationship between packet losses and simulation iterations is represented as shown in Fig. 2.

Based on Fig. 2, the conventional RSSI-based protocol has 45 as the largest packet loss and 41 as the smallest loss. However, 43 is the largest packet loss, and 13 is the smallest packet loss for the proposed scheme. On average, RSSI-based protocol has a packet loss of 49, while the proposed algorithm has an average loss of 30 packets. This represents a 38.8% reduction in packet losses for the proposed algorithm.

Computation Cost The cryptographic operations involved in typical authentication and key agreement protocol include hash operations (T_H), modular exponential (T_{ME}), elliptic curve scalar multiplication (T_{EM}), bilinear pairing (T_{BP}), signature verification (T_{SV}), and asymmetric encryption operations (T_{AE}). However, based on Algorithm 2, 2TH and $2T_{AE}$ operations are required on the UE side, while 1TH and $2T_{AE}$ operations are needed on the gNB side. As such, a total of 3TH and $4T_{AE}$ are executed in the proposed algorithm. Based on the values in [33], a single T_H and T_{AE} operations take 0.005174 ms and 0.0011 ms, respectively. Therefore, the proposed protocol has 0.019922 ms as the total computation costs as shown in Table 9.

On the other hand, the schemes in [28, 33, 34] take 0.20696 ms, 0.18626 ms, and 0.11899 ms, respectively. Consequently, the proposed algorithm had the least computation costs. As such, the proposed algorithm is efficient and applicable in 5G networks. This is because in the 5G environment application domains such as in

IoT, most of the devices such as sensors are powered via batteries, and hence, they have low computation power.

5 Conclusion and Future Work

The 5G networks have stringent latency and security requirements that have been shown to be challenging to achieve in conventional handover algorithms. Poor selection of target cells leads to deteriorating quality of services and quality of experience on the side of the users. It also leads to ping-pong handovers, which cause high latencies during the handover process. This contradicts the 5G requirements of extremely low communication latencies. Based on International Telecommunication Union recommendations, the handover latency must be below 200 ms. In addition, the control plane delay budget for X2 handover preparation and completion phases is 31 ms. For the 5G HetNets, the latency requirements are twice rigid in the control plane. The average handover latency in the proposed algorithm is 0.0318 s or 31.8 ms, which is way below the 200 ms recommendations. The proposed algorithm also offers salient security features such as resilience against replay and de-synchronization attacks. Future research directions may involve the security and performance analysis of this algorithm using other figures of merit that were not deployed in this study.

References

1. Naeem, B., Ngah, R., Hashim, S.Z.M.: Reduction in ping-pong effect in heterogeneous networks using fuzzy logic. Soft. Comput. **23**(1), 269–283 (2019)
2. Nyangaresi, V.O., Rodrigues, A.J., Abeka, S.O.: ANN-FL secure handover protocol for 5G and beyond networks. In: Zitouni, R., Phokeer, A., Chavula, J., Elmokashfi, A., Gueye, A., Benamar, N. (eds.) Towards New e-Infrastructure and e-Services for Developing Countries. AFRICOMM 2020. Lecture Notes of the Institute for Computer Sciences, Social Informatics and Telecommunications Engineering, vol. 361. Springer, Cham (2021)
3. Azzali, F., Ghazali, O., Omar, M.H.: Fuzzy Logic-based Intelligent Scheme for Enhancing QoS of Vertical Handover Decision in Vehicular Ad-hoc Networks. International Research and Innovation Summit (IRIS2017), vol. 226, pp. 1–12 (2017)
4. Pragati, K., Haridas, S.L.: Reducing Ping-Pong Effect in Heterogeneous Wireless Networks Using Machine Learning Intelligent Communication, Control and Devices, pp. 697–705 (2019)
5. Yang, H., Raza, S.M., Kim, M., Le, D.T., Van Vo, V., Choo, H.: Next point-of-attachment selection based on long short term memory model in wireless networks. In: 14th International Conference on Ubiquitous Information Management and Communication (IMCOM), pp. 1–4 (2020)
6. Nyangaresi, V.O., Abeka, S.O., Rodrigues, A.J.: Tracking area boundary-aware protocol for pseudo stochastic mobility prediction in LTE networks. Inf. Technol. Comput. Sci. **5**, 52–62 (2020)
7. Donald, E., Nosa, F.: Analyzing GSM insecurity. Int. J. Res. Sci. Innov. **3**(10), 10–18 (2016)

8. Taha, M., Parra, L., Garcia, L., Lloret, J.: An intelligent handover process algorithm in 5G networks: The use case of mobile cameras for environmental surveillance. In: Proceedings of the 2017 IEEE International Conference on Communications Workshops, ICC Workshops 2017, Paris, France, pp. 840–844 (2017)
9. Alican, O., Maode, M.: Secure and efficient vertical handover authentication for 5G HetNets. In: Proceedings of IEEE International Conference on Information Communication and Signal Processing, pp. 26–32 (2018)
10. Arshad, R., ElSawy, H., Sorour, S., Al-Naffouri, T.Y., Alouini, M.S.: Handover management in dense cellular networks: a stochastic geometry approach. In: 2016 IEEE international conference on communications (ICC), pp. 1–7. IEEE (2016)
11. Yazdinejad, A., Parizi, R.M., Dehghantanha, A., Choo, K.K.R.: Blockchain-enabled authentication handover with efficient privacy protection in SDN-based 5G networks. IEEE Trans. Netw. Sci. Eng. (2019) 1–12 (2019)
12. Hu, S., Yu, B., Qian, C., Xiao, Y., Xiong, Q., Sun, C., Gao, Y.: Non-orthogonal interleave-grid multiple access scheme for industrialinternet of things in 5G network. IEEE Trans. Indus. Inf. **14**(12), 5436–5446 (2018)
13. Cao, J., Ma, M., Li, H., Zhang, Y., Luo, Z.: A survey on security aspects for LTE and LTE-A networks. IEEE Commun. Surv. TUTs. **16**(1), 283–302 (2014)
14. Nyangaresi, V.O., Rodrigues, A.J., Abeka, S.O.: Neuro-fuzzy based handover authentication protocol for ultra dense 5G networks. In: 2020 2nd Global Power, Energy and Communication Conference (GPECOM), IEEE, pp. 339–344 (2020)
15. Bilen, T., Berk, C., Kaushik, R.C.: Handover management in software-defined ultra-dense 5G networks. IEEE Netw. **17**, 49–55 (2017)
16. Amit, K., Hari O.: Design of a USIM and ECC based handover authentication schemefor 5G-WLAN heterogeneous networks. Digital Communications and Networks, pp. 1–13 (2019)
17. Nyangaresi, V.O., Rodrigues, A.J., Abeka, S.O.: Efficient group authentication protocol for secure 5g enabled vehicular communications. In: 2020 16th International Computer Engineering Conference (ICENCO), IEEE, pp. 25–30 (2020)
18. Cheng, X., Xiaohong, H., Maode, M., Hong, B.: An anonymous handover authentication scheme based on LTE-A for vehicular networks. Wirel. Commun. Mob. Comput. **2018**, 1–16 (2018)
19. Taha, M., Jimenez, J.M., Canovas, A., Lloret, J.: Intelligent algorithm for enhancing MPEG-DASH QoE in Embms. Netw. Protoc. Algorithm. **9**(3–4), 94 (2018)
20. Phemina, M., Sendhilnathan, S.: Fuzzy based mobility management in 4G wireless networks. Braz. Arch. Biol. Technol. **59**(2), 1–13 (2017)
21. Coqueiro, T., José, J., Tássio, C., Renato, F.: A fuzzy logic system for vertical handover and maximizing battery lifetime in heterogeneous wireless multimedia networks. Wirel. Commun. Mob. Comput. **2019**, 1–14 (2019)
22. Nyangaresi, V.O., Rodrigues, A.J., Abeka, S.O.: Machine learning protocol for secure 5G handovers. Int. J. Wirel. Inf. Netw. pp. 1–22 (2022)
23. Mahira, A.G., Subhedar, M.S.: Handover decision in wireless heterogeneous networks based on feed forward artificial neural network. In: Computational Intelligence in Data Mining, pp. 663–669. Springer, Singapore (2017)
24. Benaatou, W., Latif, A., Pla, V.: Applying ANFIS model in decision-making of vertical handover between macrocell and femtocell integrated network. J. Telecommun. Electron. Comput. Eng. **11**(1), 57–62 (2019)
25. Shanmugam, K.: A novel candidate network selection based handover management with fuzzy logic in heterogeneous wireless networks. In: 4th International Conference on Advanced Computing and Communication Systems (ICACCS), IEEE, pp. 1–6 (2017)
26. Imad, E., Rachid, S., Mohammed, E.: Vertical handover decision algorithm using ants' colonies for 4G heterogeneous wireless networks. J. Comput. Netw. Commun. **2016**, 1–15 (2016)
27. Aibinu, A., Onumanyi, J., Adedigba, P., Ipinyomi, M., Folorunso, T., Salami, M.: Development of hybrid artificial intelligent based handover decision algorithm. Int. J. Eng. Sci. Technol. **20**(2), 381–390 (2017)

28. Wu, T.Y., Lee, Z., Obaidat, M.S., Kumari, S., Kumar, S., Chen, C.M.: An authenticated key exchange protocol for multi-server architecture in 5G networks. IEEE Access. **8**, 28096–28108 (2020)
29. Lai, Y., Cheng, P., Lee, C., CKu, C.: A new ticket-based authentication mechanism for fast handover in mesh network. Department of Photonics and Communication Engineering, Asia University, Taichung, Taiwan, pp. 1–18 (2016)
30. Copet, P., Marchetto, G., Sisto, R., Costa, L.: Formal Verification of LTE-UMTS Handover Procedures. IEEE, pp. 1–8 (2015)
31. Lin, Y., Longjhuang, W., Chen, Y.C.: Enhanced 4G LTE authentication and handover mechanism. Int J Electr Electron Data Commun. **3**(9), 45–47 (2015)
32. Vanga, O., Das, A.K., Goswami, A.: An efficient ECC-based privacy-preserving client authentication protocol with key agreement using smart card. J. Inf. Secur. Appl. **21**, 1–19 (2015)
33. Wu, F., Li, X., Xu, L., Sangaiah, A.K., Rodrigues, J.J.: Authentication protocol for distributed cloud computing: an explanation of the security situations for internet-of-things-enabled devices. IEEE Consum. Electron. Mag. **7**(6), 38–44 (2018)
34. Amin, R., Kumar, N., Biswas, G.P., Iqbal, R., Chang, V.: A light weight authentication protocol for IoT-enabled devices in distributed cloud computing environment. Futur. Gener. Comput. Syst. **78**, 1005–1019 (2018)
35. Mahmoud E.O., Mohamed H.M., Hassan A.: Design and simulation of a new intelligent authentication for handover over 4G (LTE) mobile communication network. In: Proceedings of the 11th ICEENG Conference, pp. 1–12 (2018)
36. Sridevi, B., Mohan, D.: Security analysis of handover key management among 4G LTE entities using device certification. Int. J. Electr. Comput. Eng. Commun. **1**(2), 1–7 (2015)
37. Nyangaresi, V.O., Abeka, S.O., Rodrigues, A.J.: Delay sensitive protocol for high availability LTE handovers. Am. J. Netw. Commun. **9**(1), 1–10 (2020)

Advances in Coyote Optimization Algorithm: Variants and Applications

Yassine Meraihi, Asma Benmessaoud Gabis, Amar Ramdane-Cherif, and Dalila Acheli

1 Introduction

In recent decades, the field of optimization has gained an increasing interest in the use of meta-heuristics for tackling real-world issues in several areas such as operation research, computer science, biology, chemistry, economics, and engineering. Meta-heuristics are generally classified into two classes: single-based and population-based meta-heuristics. In single-based meta-heuristics, one solution is obtained at each run during the process of optimization using the neighborhood process. The most popular single-based methods are: Stochastic Local Search (SLS) [1], Tabu Search (TS) [2], Variable Neighborhood Search (VNS) [3], Simulated Annealing (SA) [4], Guided Local Search (GLS) [5], and Iterated Local Search (ILS) [6]. As for population-based meta-heuristics, the main principle is the formation of a set of multiple solutions at each run. The class of population-based methods can be divided into four main categories: evolutionary, physics, swarm intelligence, and event-based. Examples of swarm-intelligence methods include Krill Herd (KH) [7], Aquila Optimizer (AO) [8], African Vulture Optimization Algorithm (AVOA) [9], Grey Wolf Optimizer (GWO) [10, 11], Bald Eagle Search

Y. Meraihi (✉)
LIST Laboratory, University of M'Hamed Bougara Boumerdes, Boumerdes, Algeria
e-mail: y.meraihi@univ-boumerdes.dz

A. B. Gabis
Ecole nationale Supérieure d'Informatique, Laboratoire des Méthodes de Conception des Systémes, Oued-Smar, Alger, Algeria

A. Ramdane-Cherif
LISV Laboratory, University of Versailles St-Quentin-en-Yvelines, Velizy, France

D. Acheli
LAA Laboratory, University of M'Hamed Bougara Boumerdes, Boumerdes, Algeria

M. D. Hina et al. (eds.), *Advances in Computational Intelligence and Communication*, EAI/Springer Innovations in Communication and Computing,
https://doi.org/10.1007/978-3-031-19523-5_7

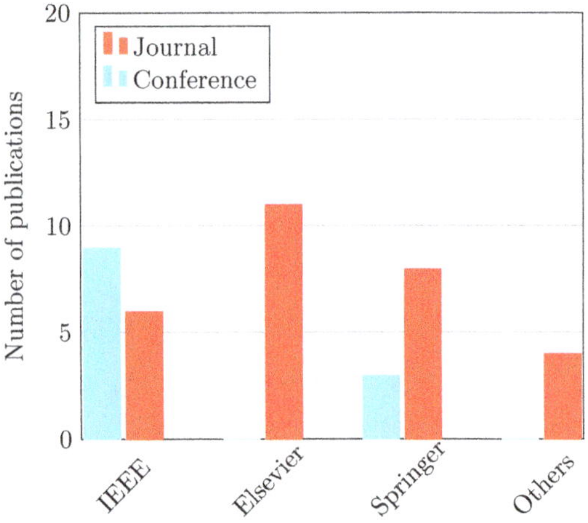

Fig. 1 Number of WMNs node placement related publications per database

optimization (BES) [12], Fruit Fly Optimization (FFO) [13], Dragonfly Algorithm (DA) [14, 15], Ant Lion Optimizer (ALO) [16], Particle Swarm Optimization (PSO) [17], Dolphin Echolocation Algorithm (DEA) [18], Crow Search Algorithm (CSA) [19, 20], Grasshopper Optimization Algorithm (GOA) [21, 22], Glowworm Swarm Optimization (GSO) [23], Salp Swarm Algorithm (SSA) [24], Great Salmon Run Algorithm (GSRA) [25], Monkey Search Algorithm [26], and Coyote Optimization Algorithm (COA) [27].

The COA population-based algorithm was introduced by Pierezan et al. [27] in 2018 for tackling global optimization issues. COA was assessed based on 40 famous benchmark functions with different classifications. It has been proven that COA provides good solutions in a reasonable time.

This paper gives a brief overview of COA, its variants (modified and hybridized versions) and its utilization in different fields. We have analyzed a large number of papers considering various popular publishers (i.e., IEEE, Elsevier, Springer, and others). Figure 1 presents the number of related COA papers by publishers where IEEE is the one who produce the most about COA. Figure 2 presents the number of COA publications by year.

The remaining of this review is organized as follows. Section 2 gives the structure of the original COA. Section 3 gives the variants of COA. Section 4 presents the applications of COA in various fields. Finally, Sect. 5 provides the concluding remarks.

2 Coyote Optimization Algorithm (COA)

Pierezan and Coelho [27] developed in 2018 a newly swarm-intelligence technique, called Coyote Optimization Algorithm (COA), for tackling global optimization

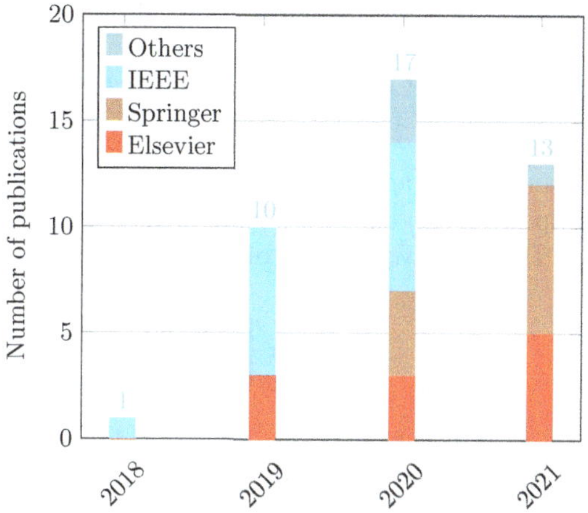

Fig. 2 Number of COA publications per year

issues. COA algorithm is inspired by the attitude of the Canis Latrans species in nature [27]. Coyotes are divided into G_p classes with C_c coyotes in each class. The population size can be given by $G_p{*}C_c$. The social condition of the coyote c in the group p at iteration t can be given as follows [27]:

$$SC_c^{p,t} = X = (x_1, x_2, \ldots, x_D) \tag{1}$$

where D is the dimension of the search space.

This attitude engages adapting coyotes to the environment named $ft_c^{p,t} \in R$. The adjustment of the coyote to its corresponding current condition is given as follows:

$$ft_c^{p,t} = f(SC_c^{p,t}) \tag{2}$$

The coyotes are randomly affected to the groups at the beginning of the process. In each group, a leader, named the alpha coyote, is selected. It represents the coyote that obtain the best fitness. The alpha coyote can be given as follows:

$$\text{alpha}^{p,t} = \{SC_c^{p,t} | \arg_{c=\{1,2,\ldots,C_c\}} \min(\max) f(SC_c^{p,t})\} \tag{3}$$

On its side, the cultural tendency of each group is given as follows

$$\text{cultr}_j^{p,t} = \begin{cases} R^{p,t}_{\frac{Cc+1}{2},j}, & Cc \quad is \quad \text{odd} \\ \frac{R^{p,t}_{\frac{Cc}{2},j} + R^{p,t}_{\frac{Cc+1}{2},j}}{2}, & \text{Otherwise} \end{cases} \tag{4}$$

where $R^{p,t}$ is the ranked social condition of the coyote inside the group p.

The new social conditions of the different coyotes are updated as follows:

$$\text{new\_}SC_c^{p,t} = \left\{SC_c^{p,t} + r_1(\text{alpha}^{p,t} - SC_{cr_1}^{p,t}) + r_2(\text{cultr}^{p,t} - SC_{cr_2}^{p,t}\right\} \quad (5)$$

where $SC_{cr_1}^{p,t}$ and $SC_{cr_2}^{p,t}$ are social conditions of random coyotes cr_1 and cr_2, respectively. r_1 and r_2 are random numbers in [0, 1].

The objective function of the new social condition is expressed as follows:

$$ft_c^{p,t} = \begin{cases} \text{new\_}ft_c^{p,t}, & \text{if new\_}ft_c^{p,t} > ft_c^{p,t} \\ ft_c^{p,t}, & \text{Otherwise} \end{cases} \quad (6)$$

The final social conditions are selected based on the computation of new solutions. Each new solution is compared with the worst one in each group. The best solution is chosen as follows:

$$SC_c^{p,t} = \begin{cases} \text{new\_}SC_c^{p,t}, & \text{if new\_}ft_c^{p,t} > ft_c^{p,t} \\ SC_c^{p,t}, & \text{Otherwise} \end{cases} \quad (7)$$

The pseudo-code of COA is illustrated in Algorithm 1 [27]

Algorithm 1 The pseudo-code of COA

Initialize randomly the social conditions of G_p groups with C_c coyotes each
Assess the coyotes adaptation (Eq. 2)
while stopping criterion is not achieved **do**
  for p = 1 to G_p (all G_p groups in the population) **do**
    Generate the alpha coyote of the group p (Eq. 4)
    Calculate the cultural tendency of the group p (Eq. 5)
    for c = 1 to C_c (all C_c coyotes) **do**
      Generate the new social conditions (Eq. 6)
      Evaluate the new social conditions (new fitness) (Eq. 7)
      Check the adaptation of the social conditions (Eq. 8)
    end for
    Generate a new coyote by combining two random parents (Eq. 9)
  end for
  Transition of random coyotes between random group (Eq. 3)
  Update the ages of all coyotes
end while
Return the best social conditions

3 Recent Variants of Coyote Optimization Algorithm

A number of recent variants of the original COA have been proposed in the literature, classified into modified and combined versions. The details of these versions are given below.

3.1 Modified Coyote Optimization Algorithm

Some modified versions of COA were introduced by investigating new strategies such as chaotic map, Levy flight, and local search. The detail of each modified version is given in the following.

Pierezan et al. [28] developed an ameliorated COA (MCOA) considering the chaotic Tinkerbell map scheme for tackling the truss optimization problems. The considered MCOA was validated using four benchmark problems and results revealed that MCOA provides competitive solutions compared to other optimization approaches.

In the work of Tong et al. [29], a chaotic COA (CCOA) was developed for global optimization problems. Ten chaotic maps were investigated to tune some parameters of COA. CCOA was evaluated based on ten well-known benchmark functions and results approved that CCOA outperforms COA in terms of rapid global convergence rate.

Dangi et al. [30] developed a chaotic COA technique based on Tinkerbell map for sentiment analysis of social media data. Comparative results demonstrated that the proposed chaotic COA outperforms some well-known existing approaches.

Kaymaz et al. [31] introduced an ameliorated COA (LCOA) by integrating the Levy flight strategy into COA to deal with the optimal power flow issue. LCOA was tested based on three test systems considering five parameters (i.e., fuel cost, voltage stability, emissions, voltage profile, and active power loss). Simulation results showed that COA is more effective and robust compared to other optimization techniques.

Duman et al. [32] introduced an enhanced COA, called LRFDBCOA, by incorporating Lévy flight and fitness-distance balance (FDB) strategies into COA for tackling global optimization and optimal power flow problems. The suggested LRFDBCOA was tested with 28 other meta-heuristic approaches and results proved its successfulness in tackling the optimal power flow issue.

Huang and Zhuang [33] developed an ameliorated COA by integrating the Sinusoidal map concept and Lévy flight mechanism for optimizing the Pitch Control parameters of Wind Turbine. The proposed solution was approved considering multi-step and IEC turbulence wind conditions and results showed its robustness and effectiveness.

Wu et al. [34] developed an ameliorated COA (ICOA) based on sobol sequence strategy for optimal load forecasting of ELM model. The effectiveness of the ameliorated ICOA was assessed based 23 well-studied benchmark functions and real data of the historical load in Zhengzhou from 2016 to 2017. Obtained results approved the robustness of ICOA compared to other optimizer models.

Nguyen et al. [35] proposed an improved COA (ICOA) to overcome the electric distribution network reconfiguration problem. In ICOA, a local search strategy was proposed to ameliorate the so-far best solution. The robustness of ICOA was evaluated based on three systems (i.e., 25 node, 33 node, and 69 node). According to results, ICOA outperformed the original COA, SFO, PSO, and other optimization approaches.

de Souza et al. [36] proposed a binary COA (BCOA) for the optimal selection of feature subset for classification. The suggested BCOA was assessed considering the computational cost, accuracy, and classification and numerical results showed that BCOA achieves good performance in comparison with some optimization models in the literature.

Yuan et al. [37] introduced a developed COA (DCOA) for ideal parameter measurement of a proton-exchange membrane fuel cell model. Authors considered two models such as A2 kW Nexa FC and 6kW NedSstack PS6 FC to assess the performance of DCOA. According to results, DCOA showed excellence performance towards the original COA and other techniques found in the literature.

Shi et al. [38] developed an ameliorated COA (ICOA) for ideal configuration of photovoltaic intelligent edge terminal. A single-dimension growth strategy was introduced to prevent premature convergence. Numerical results validated that ICOA gives better performance than COA in terms of accuracy and stability.

Nguyen et al. [39] proposed an improved COA (ICOA) for perfect position and size of solar photovoltaic distribution generation units. The effectiveness of ICOA was tested using two power test systems such as IEEE 33-bus and IEEE 69-bus. Obtained results approved the efficiency of ICOA compared to COA, SFO, SSA, and GA in terms of stability and convergence rate.

Pham et al. [40] introduced an ameliorated COA (ECOA) for optimal location and capacity of distributed generations. The effectiveness of ECOA was assessed based on IEEE 85-bus, IEEE 69-bus, and IEEE 33-bus radial distribution networks. Obtained results approved that the ECOA gives good performance compared to SFO, SSA, and COA techniques.

Alghamdi [41] proposed an improved COA (ICOA) for optimum location of DGs in a Power System. The effectiveness of ICOA was approved using two test distribution system such as IEEE 21-bus and 28-bus. Obtained results demonstrated that ICOA exceeds other optimizer techniques in selection best locations of DGs in distribution networks.

Abou El-Ela et al. [42] developed an ameliorated COA (ECAO) for optimizing the parameters of PDn-PI controller in multi-area power systems. The robustness of ECOA was assessed based on various sets of disturbances and results proved the successfulness and robustness of the suggested ECOA in comparison with other models found in the literature.

3.2 Hybridized Coyote Optimization Algorithm

Some combined techniques based on combining COA with other optimization algorithms such as cultural algorithm, fuzzy logic model, and support vector regression have been developed. Details of these combined techniques are given below.

Amin et al. [43] proposed a hybrid approach by hybridizing the analytical technique with COA for optimal position of distribution static compensators. The efficacy of the hybrid technique was validated using IEEE 33-bus and 69-bus test systems and results showed its good performance in comparison with some techniques.

Pierezan et al. [44] developed an new method, called Cultural COA (CCOA), based on hybridizing COA with Cultural Algorithm (CA) for tackling the gas turbine issue. CCOA was tested based on five case studies considering physical and mechanical constraints. Experimental results revealed that CCOA can enhance significantly the gas turbine operation and reduce fuel consumption up to 3.6%.

Jin et al. [45] developed a hybrid COA with GWO for adjusting the control parameters of SMDTC. The proposed solution was evaluated by experiments and results approve it efficiency and effectiveness.

Kumar et al. [46] developed a combined model by hybridizing COA with GWO and Deep CNN models for arrhythmia classification. The combined model was approved based on MIT-BIH Database and results demonstrated its robustness compared to other classifier models by achieving up to 95% of accuracy.

Abdelghafar et al. [47] proposed a hybrid model, called COA-SVR, based on the hybridization of COA with SVR for estimating the battery remaining useful life. The effectiveness of COA-SVR was evaluated using NASA's Lithium ion batteries dataset and results proved that the considered method achieves good performance in terms of stability and time complexity.

Mohamed et al. [48] proposed a hybrid approach, called COFL, combining COA with Fuzzy logic scheme for energy efficiency in WSNs. The performance of COFL was tested based on three different scenarios considering the metrics of wireless network lifetime, energy consumption, and throughput. Simulation results showed that COFL exceeds other optimization models such as traditional COA, LEACH protocol, and SEP model. Similarly, Abaza et al. [49] developed a new model (ECOA) based on combining COA with fuzzy logic principles for tackling the reactive power dispatch issue. The suggested ECOA was assessed based on three test systems (i.e., IEEE 30-bus, 118-bus, and IEEE 300-bus) and results proved that ECOA gives competitive solutions by reducing significantly the power losses.

Li et al. [50] proposed a hybrid technique (FICOA) combining COA with Fuzzy logic model and differential evolution for multilevel image thresholding. The efficiency of FICOA was tested based on data sets taken from Berkeley Segmentation Data Set BSD500 considering PSNR and FSIM parameters. Experimental results validated that FICOA achieves better image segmentation quality compared to some optimization techniques.

4 Application of the Coyote Optimization Algorithm

COA has been used for solving various optimization issues in several domains such as image segmentation, feature selection, economic load dispatch, maximum power point tracking, distributed generations in radial distribution networks, and electric power system. The details are provided below.

4.1 Image Segmentation

In the work of Moses [51], COA was applied for image segmentation. The performance of COA was approved based on 40 benchmark functions with different characteristics taken from the IEEE CEC2015. Test results showed the superiority of COA compared to other techniques by obtaining accurate threshold values for image segmentation.

Sayed et al. [52] used multi-swarm COA (MCOA) for automatic skin lesion segmentation. The multi-swarm scheme was applied where the population is categorized into different sub-classes. The efficiency of MCOA was tested using a set of numerical benchmark functions. Simulation results illustrate the robustness of MCOA compared to the COA algorithm and other existing models.

4.2 Economic Load Dispatch

Güvenç and Kaymez [53] used COA to tackle the economic dispatch integrated wind power issue. COA was assessed using 6 and 12 generator test systems consisting of wind power generators and traditional thermal generators. Numerical results illustrated the efficiency of COA compared to PSO and GA algorithms.

Güvenç and Battal [54] used COA for solving the energy hub economic dispatch issue. The performance of COA was tested considering three parameters (i.e., emission, cost, security of multiple energy carriers and structures). Obtained results revealed that COA performs better than other techniques such as MSA, GA, SOS, and PSO models.

4.3 Distributed Generations in Radial Distribution Networks

In [55], COA was applied for the optimal position of distributed generations. The considered COA was assessed based on IEEE 69-bus system to decrease the power losses and ameliorate the voltage profile under operation constraints. Numerical results validated the efficiency and accuracy of COA compared to well-regarded methods.

Abdallah et al. [56] used COA for optimal sizing and position of renewable distributed generations. The efficiency of COA was tested using two IEEE test systems and results proved its feasibility to obtain optimum position and size renewable distributed generations.

Chang et al. [57] used COA for optimal planning of photovoltaic distributed generations (PV-DG). COA was assessed using the IEEE 123-bus benchmark system and an actual utility distribution feeder in comparison with GA, BBO, PSO, and GWO. Simulation results proved the successfulness of COA compared to other methods in terms of hosting capacity and power loss reduction for PV-DG planning.

Nguyen et al. [58] employed COA for tackling the issue of simultaneous network reconfiguration and distributed generation position. The performance of COA was approved based on two distribution systems (i.e., 69_node and 119_node) and results showed that COA gives the better network configuration and position of DGs compared to other methods.

Janamala and Reddy [59] applied COA for optimal allocation of the interline-photovoltaic battery storage system. The robustness of COA was validated using the IEEE 33-bus electrical distribution network and results proved its successfulness compared to GOA and PSO methods.

Abdallah et al. [60] used COA for optimal position and size of renewable distributed generations in radial distribution networks. COA was approved using IEEE 33-bus, IEEE 69-bus systems and results showed its performance.

4.4 Optimal Parameters Control

Babu and Saikia [61] employed COA for adjusting the parameters of PI minus DF controller for LFC in multi-area power systems. COA was evaluated using fixed and random insolation among different areas and obtained results showed the successfulness of the considered ECOA for optimal PI minus DF controller of LFC.

Moschos and Parisses [62] used COA for obtaining good parameters of the PIkDND2N2 controller for an Automatic Voltage Regulator (AVR). COA was evaluated using the transient response and disturbance rejection metrics. Simulation results validated the successfulness of the COA-based PIkDND2N2 controller compared to other existing controller models.

Guesmi et al. [63] used COA for adjusting the parameters of PSSs and SVC in Multi-machine Power System. COA was assessed considering the 3-machine 9-bus WSCC system and results approved the robustness of COA in comparison to the GA and PSO models.

4.5 Parameters Estimation of Electrical Models

Qais et al. [64] applied COA for unknown parameters extraction of the three-diode photovoltaic model of photovoltaic modules. The effectiveness of COA was evaluated using certain commercial photovoltaic modules in the market such as the KC200GT and MSX-60 modules. Experimental results revealed that using COA, a highly accurate three-diode photovoltaic model is achieved.

Chin and Salam [65] used COA for parameters extraction of single diode and two-diode photovoltaic cells/models. The efficiency of COA was experimented with three photovoltaic modules (i.e., thin-film, mono-crystalline, and multi-crystalline) and results showed that COA gives more accurate solutions than some optimization models in the literature.

Abaza et al. [66] used COA for accurate parameter measurement of solid oxide fuel cell model. The robustness of COA was tested using 5-kW SOFC stack under various operating situations in comparison with the ranking teaching-learning optimizer (RTLBO) technique. Simulation results revealed the successfulness of COA compared to RTLBO technique.

Diab et al. [67] employed COA for parameters measurement of single diode, two-diode, and three-diode for photovoltaic cells/models. The robustness of COA was assessed using three types of photovoltaic modules (i.e., thin-film, mono-crystalline, and multi-crystalline). Simulation results revealed the successfulness of COA for parameter estimation of different photovoltaic modules.

Abdelwanis et al. [68] used COA for the parameters measurement of electrical power transformers. The capability of COA was tested based on two cases (i.e., single-phase transformer and three-phase transformer). Obtained results showed the successfulness of COA compared to JAYA and PSO models.

Sultan et al. [69] used COA for the extraction of unknown parameters of proton-exchange membrane fuel cell models. The effectiveness of COA was evaluated under different operating conditions using seven and ten unknown parameters. Simulation results proved that COA provides competitive solutions compared to other recent optimization techniques reported in the literature.

4.6 Maximum Power Point Tracking (MPT)

Mostafa et al. [70] applied COA to track the global MPT of the photovoltaic system considering the effect of partial shading. The efficiency of COA was tested in comparison with enhanced GOA (E-GWO), DA, PSO, and ALO. Results approved that COA exceeds other methods in terms of tracking capability, steady-state, and efficiency.

In another similar work, Pervez et al. [71] used COA for maximum MPT of a partially shaded solar photovoltaic generation system. The performance of COA was evaluated using Canadian Solar CS6X 305-M PV module for different insolation conditions. Results approved the superiority of COA in comparison with GSA and PSO techniques.

4.7 Other Applications

Mouhamed et al. [72] applied COA for watermarking 3D printing data. The efficiency of COA was validate based on a set of 3D mesh models used in most of watermark techniques. Results approved the successfulness of COA compared to the most known watermarking methods.

Rezk et al. [73] used COA for tackling the reconfiguration mechanism of the partially shaded photovoltaic array. The performance of COA was tested using 9×9 photovoltaic array under four different shadow patterns. Results revealed the competence of COA compared to other techniques found in the literature.

Fathy et al. [74] employed COA to ameliorate the fuel economy of hybrid fuel cell, super-capacitors, and battery storage bank systems. The validity and effectiveness of COA was investigated based on high energy efficiency and minimum hydrogen consumption in comparison with EEMS, GA, GWO, GOA, MVO, SSA, and PSO. Simulation results showed that COA reduces hydrogen consumption by 38.8% and outperforms some well-regarded optimization approaches.

Babu et al. [75] employed COA for automatic load frequency control of an unequal three-area thermal system. The superiority of COA was evaluated in comparison with FA and CS in terms of convergence characteristics and results revealed its good performance for automatic generation control studies.

5 Conclusion

COA is a promising population-based meta-heuristic that has attracted increasing attention from scientists and researchers since it was proposed in 2018. This work gives the first literature review of COA according to its variants (hybridizations and modifications) and applications. We reveal that related works proved the performance and efficiency of COA to provide good solutions in a reasonable execution time. However, there are still many propositions for future works. In the area of COA hybridization, combining COA with some meta-heuristics (i.e., GA, ACO, PSO, DA, FA, GWO, KH, CS, CSO, MFO, and CSA) are needed to ameliorate both intensification and diversification. Another possible area of variants COA is the development of improved versions of COA by investigating new schemes and operators such as Gaussian, Mutation, Opposition-based, and Orthogonal-based strategies. Finally, COA can be applied to handle other practical problems such as image processing, scheduling, multicast routing problem, machine learning, robotics, and other engineering optimization problems.

References

1. Hoos, H.H., Stützle, T.: Stochastic Local Search: Foundations and Applications. Elsevier, Amsterdam (2004)
2. Glover, F., Laguna, M.: Tabu search. In: Handbook of Combinatorial Optimization, pp. 2093–2229. Springer, Berlin (1998)
3. Mladenović, N., Hansen, P.: Variable neighborhood search. Comput. Oper. Res. **24**(11), 1097–1100 (1997)
4. Kirkpatrick, S., Gelatt, C.D., Vecchi, M.P.: Optimization by simulated annealing. Science **220**(4598), 671–680 (1983)
5. Voudouris, C., Tsang, E.: Guided local search and its application to the traveling salesman problem. Eur. J. Oper. Res. **113**(2), 469–499 (1999)
6. Lourenço, H.R., Martin, O.C., Stützle, T.: Iterated local search. In: Handbook of Metaheuristics, pp. 320–353. Springer, Berlin (2003)
7. Gandomi, A.H., Alavi, A.H.: Krill herd: a new bio-inspired optimization algorithm. Commun. Nonlinear Sci. Numer. Simul. **17**(12), 4831–4845 (2012)
8. Abualigah, L., Yousri, D., Elaziz, M.A., Ewees, A.A., Al-qaness, M.A.A., Gandomi, A.H.: Aquila optimizer: a novel meta-heuristic optimization algorithm. Comput. Ind. Eng. **157**, 107250 (2021)
9. Abdollahzadeh, B., Gharehchopogh, F.S., Mirjalili, S.: African vultures optimization algorithm: a new nature-inspired metaheuristic algorithm for global optimization problems. Comput. Ind. Eng. **158**, 107408 (2021)
10. Mirjalili, S., Mirjalili, S.M., Lewis, A.: Grey wolf optimizer. Adv. Eng. Softw. **69**, 46–61 (2014)
11. Faris, H., Aljarah, I., Al-Betar, M.A., Mirjalili, S.: Grey wolf optimizer: a review of recent variants and applications. Neural Comput. Applic. **30**(2), 413–435 (2018)
12. Alsattar, H.A., Zaidan, A.A., Zaidan, B.B.: Novel meta-heuristic bald eagle search optimisation algorithm. Artif. Intell. Rev. **53**(3), 2237–2264 (2020)
13. Pan, W.-T.: A new fruit fly optimization algorithm: taking the financial distress model as an example. Knowl. Based Syst. **26**, 69–74 (2012)
14. Mirjalili, S.: Dragonfly algorithm: a new meta-heuristic optimization technique for solving single-objective, discrete, and multi-objective problems. Neural Comput. Applic. **27**(4), 1053–1073 (2016)
15. Meraihi, Y., Ramdane-Cherif, A., Acheli, D., Mahseur, M.: Dragonfly algorithm: a comprehensive review and applications. Neural Computi. Applic. **32**(21), 16625–16646 (2020)
16. Mirjalili, S.: The ant lion optimizer. Adv. Eng. Softw. **83**, 80–98 (2015)
17. Eberhart, R., Kennedy, J.: A new optimizer using particle swarm theory. In: MHS'95. Proceedings of the Sixth International Symposium on Micro Machine and Human Science, pp. 39–43. IEEE, Piscataway (1995)
18. Kaveh, A., Farhoudi, N.: A new optimization method: Dolphin echolocation. Adv. Eng. Softw. **59**, 53–70 (2013)
19. Askarzadeh, A.: A novel metaheuristic method for solving constrained engineering optimization problems: crow search algorithm. Comput. Struct. **169**, 1–12 (2016)
20. Meraihi, Y., Gabis, A.B., Ramdane-Cherif, A., Acheli, D.: A comprehensive survey of crow search algorithm and its applications. Artif. Intell. Rev., 1–48 (2020)
21. Saremi, S., Mirjalili, S., Lewis, A.: Grasshopper optimisation algorithm: theory and application. Adv. Eng. Softw. **105**, 30–47 (2017)
22. Meraihi, Y., Gabis, A.B., Mirjalili, S., Ramdane-Cherif, A.: Grasshopper optimization algorithm: theory, variants, and applications. IEEE Access **9**, 50001–50024 (2021)
23. Marinaki, M., Marinakis, Y.: A glowworm swarm optimization algorithm for the vehicle routing problem with stochastic demands. Expert Syst. Applic. **46**, 145–163 (2016)
24. Mirjalili, S., Gandomi, A.H., Mirjalili, S.Z., Saremi, S., Faris, H., Mirjalili, S.M.: Salp swarm algorithm: a bio-inspired optimizer for engineering design problems. Adv. Eng. Softw. **114**, 163–191 (2017)

25. Mozaffari, A., Fathi, A., Behzadipour, S.: The great salmon run: a novel bio-inspired algorithm for artificial system design and optimisation. Int. J. Bio-Inspired Comput. **4**(5), 286–301 (2012)
26. Mucherino, A., Seref, O.: Monkey search: a novel metaheuristic search for global optimization. In: AIP Conference Proceedings, vol. 953, pp. 162–173. American Institute of Physics, College Park (2007)
27. Pierezan, J., Coelho, L.D.S.: Coyote optimization algorithm: a new metaheuristic for global optimization problems. In: 2018 IEEE Congress on Evolutionary Computation (CEC), pp. 1–8. IEEE, Piscataway (2018)
28. Pierezan, J., dos Santos Coelho, L., Mariani, V.C., de Vasconcelos Segundo, E.H., Prayogo, D.: Chaotic coyote algorithm applied to truss optimization problems. Comput. Struct. **242**, 106353 (2021)
29. Tong, H., Zhu, Y., Pierezan, J., Xu, Y., dos Santos Coelho, L.: Chaotic coyote optimization algorithm. J. Ambient. Intell. Humaniz. Comput., 1–21 (2021)
30. Dangi, D., Bhagat, A., Dixit, D.K.: Sentiment analysis of social media data based on chaotic coyote optimization algorithm based time weight-AdaBoost support vector machine approach. Concurr. Comput. Pract. Exp. **34**(3), e6581 (2022)
31. Kaymaz, E., Duman, S., Guvenc, U.: Optimal power flow solution with stochastic wind power using the lévy coyote optimization algorithm. Neural Comput. Applic. **33**(12), 6775–6804 (2021)
32. Duman, S., Kahraman, H.T., Guvenc, U., Aras, S.: Development of a lévy flight and FDB-based coyote optimization algorithm for global optimization and real-world ACOPF problems. Soft Comput. **25**(8), 6577–6617 (2021)
33. Huang, C., Zhuang, J.: Error-based active disturbance rejection control for pitch control of wind turbine by improved coyote optimization algorithm. IEEE Trans. Energy Convers. **31**(2), 1394–1405 (2021)
34. Wu, S., Jiang, J., Yan, Y., Bao, W., Shi, Y.: Improved coyote algorithm and application to optimal load forecasting model. Alex. Eng. J. **61**(10), 7811–7822 (2022)
35. Nguyen, T.T., Nguyen, Q.T., Nguyen, T.T.: Optimal radial topology of electric unbalanced and balanced distribution system using improved coyote optimization algorithm for power loss reduction. Neural Comput. Applic., 1–28 (2021)
36. Thom de Souza, R.C., de Macedo, C.A., dos Santos Coelho, L., Pierezan, J., Mariani, V.C.: Binary coyote optimization algorithm for feature selection. Pattern Recogn. **107**, 107470 (2020)
37. Yuan, Z., Wang, W., Wang, H., Yildizbasi, A.: Developed coyote optimization algorithm and its application to optimal parameters estimation of PEMFC model. Energ. Rep. **6**, 1106–1117 (2020)
38. Shi, S., Zhou, S., Zhang, L.: Application of improved coyote optimization algorithm in optimal configuration of photovoltaic intelligent edge terminal. In: 2020 IEEE 4th Conference on Energy Internet and Energy System Integration (EI2), pp. 3998–4003. IEEE, Piscataway (2020)
39. Nguyen, T.T., Pham, T.D., Kien, L.C., Dai, L.V.: Improved coyote optimization algorithm for optimally installing solar photovoltaic distribution generation units in radial distribution power systems. Complexity **2020** (2020)
40. Pham, T.D., Nguyen, T.T., Dinh, B.H.: Find optimal capacity and location of distributed generation units in radial distribution networks by using enhanced coyote optimization algorithm. Neural Comput. Applic. **33**(9), 4343–4371 (2021)
41. Alghamdi, H.: Optimum placement of distribution generation units in power system with fault current limiters using improved coyote optimization algorithm. Entropy **23**(6), 655 (2021)
42. El-Ela, A.A.A., El-Sehiemy, R.A., Shaheen, A.M., El-Gelil Diab, A.: Enhanced coyote optimizer-based cascaded load frequency controllers in multi-area power systems with renewable. Neural Computi. Applic. **33**(14), 8459–8477 (2021)
43. Amin, A., Kamel, S., Selim, A., Nasrat, L.: Optimal placement of distribution static compensators in radial distribution systems using hybrid analytical-coyote optimization technique. In: 2019 21st International Middle East Power Systems Conference (MEPCON), pp. 982–987. IEEE, Piscataway (2019)

44. Pierezan, J., Maidl, G., Yamao, E.M., dos Santos Coelho, L., Mariani, V.C.: Cultural coyote optimization algorithm applied to a heavy duty gas turbine operation. Energy Convers. Manage. **199**, 111932 (2019)
45. Jin, Z., Sun, X., Lei, G., Guo, Y., Zhu, J.: Sliding mode direct torque control of SPMSMS based on a hybrid wolf optimization algorithm. IEEE Trans. Ind. Electron. **69**(5), 4534–4544 (2021)
46. Kumar, A., Kumar, S.A., Dutt, V., Dubey, A.K., García-Díaz, V.: IOT-based ECG monitoring for arrhythmia classification using coyote grey wolf optimization-based deep learning CNN classifier. Biomed. Sig. Process. Control **76**, 103638 (2022)
47. Abdelghafar, S., Goda, E., Darwish, A., Hassanien, A.E.: Satellite lithium-ion battery remaining useful life estimation by coyote optimization algorithm. In: 2019 Ninth International Conference on Intelligent Computing and Information Systems (ICICIS), pp. 124–129. IEEE, Piscataway (2019)
48. Mohamed, A., Saber, W., Elnahry, I., Hassanien, A.E.: Coyote optimization based on a fuzzy logic algorithm for energy-efficiency in wireless sensor networks. IEEE Access **8**, 185816–185829 (2020)
49. Abaza, A., Fawzy, A., El-Sehiemy, R.A., Alghamdi, A.S., Kamel, S.: Sensitive reactive power dispatch solution accomplished with renewable energy allocation using an enhanced coyote optimization algorithm. Ain Shams Eng. J. **12**(2), 1723–1739 (2021)
50. Li, L., Sun, L., Xue, Y., Li, S., Huang, X., Mansour, R.F.: Fuzzy multilevel image thresholding based on improved coyote optimization algorithm. IEEE Access **9**, 33595–33607 (2021)
51. Moses, M.L.: Coyote optimization algorithm based multilevel thresholding approach for image segmentation. J. Soft Comput. Eng. Applic. **1**(1) (2020)
52. Sayed, G.I., Khoriba, G., Haggag, M.H.: The novel multi-swarm coyote optimization algorithm for automatic skin lesion segmentation. Evol. Intell., 1–33 (2020)
53. Güvenç, U., Kaymaz, E.: Economic dispatch integrated wind power using coyote optimization algorithm. In: 2019 7th International Istanbul Smart Grids and Cities Congress and Fair (ICSG), pp. 179–183. IEEE, Piscataway (2019)
54. Güvenç, U., Battal, O.: Coyote optimization algorithm to solve energy hub economic dispatch problem. Uluslararası Teknolojik Bilimler Dergisi **12**(1), 20–26 (2020)
55. Kamel, S., Amin, A., Selim, A., Ahmed, M.H.: Application of coyote optimizer for optimal dg placement in radial distribution systems. In: 2019 International Conference on Computer, Control, Electrical, and Electronics Engineering (ICCCEEE), pp. 1–6. IEEE, Piscataway (2019)
56. Abdallah, E.M., ELsayed, M.I., ELgazzer, M.M., Hassan, A.A.: Optimal location and sizing of renewable distributed generators in radial distribution system using coyote optimization algorithm. Tech (2020)
57. Chang, G.W., Chinh, N.C.: Coyote optimization algorithm-based approach for strategic planning of photovoltaic distributed generation. IEEE Access **8**, 36180–36190 (2020)
58. Nguyen, T.T., Nguyen, T.T., Nguyen, N.A., Duong, T.L.: A novel method based on coyote algorithm for simultaneous network reconfiguration and distribution generation placement. Ain Shams Eng. J. **12**(1), 665–676 (2021)
59. Janamala, V., Reddy, D.S.: Coyote optimization algorithm for optimal allocation of interline–photovoltaic battery storage system in islanded electrical distribution network considering EV load penetration. J. Energ. Storage **41**, 102981 (2021)
60. Abdallah, E.M., El Sayed, M.I., Elgazzar, M.M., Hassan, A.A.: Coyote multi-objective optimization algorithm for optimal location and sizing of renewable distributed generators. Int. J. Elect. Comput. Eng. **11**(2), 975 (2021)
61. Babu, N.R., Saikia, L.C.: Load frequency control of a multi-area system incorporating dish-stirling solar thermal system and coyote optimized pi minus DF controller. In: 2020 IEEE International Conference on Power Electronics, Smart Grid and Renewable Energy (PESGRE2020), pp. 1–6. IEEE, Piscataway (2020)
62. Moschos, I., Parisses, C.: A novel optimal PIλDND2N2 controller using coyote optimization algorithm for an AVR system. Eng. Sci. Technol. Int. J. **26**, 1–12 (2021)

63. Guesmi, T., Alshammari, B. M., Almalaq, Y., Alateeq, A., Alqunun, K.: New coordinated tuning of SVC and PSSs in multimachine power system using coyote optimization algorithm. Sustainability, **13**(6), 3131 (2021).
64. Qais, M.H., Hasanien, H.M., Alghuwainem, S., Nouh, A.S.: Coyote optimization algorithm for parameters extraction of three-diode photovoltaic models of photovoltaic modules. Energy **187**, 116001 (2019)
65. Chin, V.J., Salam, Z.: Coyote optimization algorithm for the parameter extraction of photovoltaic cells. Solar Energ. **194**, 656–670 (2019)
66. Abaza, A., El Sehiemy, R.A., Bayoumi, A.S.A.: Optimal parameter estimation of solid oxide fuel cell model using coyote optimization algorithm. In: Recent Advances in Engineering Mathematics and Physics, pp. 135–149. Springer, Berlin (2020)
67. Diab, A.A.Z., Sultan, H.M., Do, T.D., Kamel, O.M., Mossa, M.A.: Coyote optimization algorithm for parameters estimation of various models of solar cells and PV modules. IEEE Access **8**, 111102–111140 (2020)
68. Abdelwanis, M.I., Abaza, A., El-Sehiemy, R.A., Ibrahim, M.N., Rezk, H.: Parameter estimation of electric power transformers using coyote optimization algorithm with experimental verification. IEEE Access **8**, 50036–50044 (2020)
69. Sultan, H.M., Menesy, A.S., Kamel, S., Jurado, F.: Developing the coyote optimization algorithm for extracting parameters of proton-exchange membrane fuel cell models. Elect. Eng. **103**(1), 563–577 (2021)
70. Mostafa, H.H., Ibrahim, A.M.: Performance investigation for tracking GMPP of photovoltaic system under partial shading condition using coyote algorithm. In: 2019 21st International Middle East Power Systems Conference (MEPCON), pp. 34–40. IEEE, Piscataway (2019)
71. Pervez, I., Sarwar, A., Pervez, A., Tariq, M., Zaid, M.: Maximum power point tracking of a partially shaded solar PV generation system using coyote optimization algorithm (COA). In: Advances in Electromechanical Technologies, pp. 509–518. Springer, Berlin (2021)
72. Mouhamed, M.R., Soliman, M.M., Darwish, A., Hassanien, A.E.: Watermarking 3d printing data based on coyote optimization algorithm. In: Machine Learning and Big Data Analytics Paradigms: Analysis, Applications and Challenges, pp. 603–624. Springer, Berlin (2021)
73. Rezk, H., Fathy, A., Aly, M.: A robust photovoltaic array reconfiguration strategy based on coyote optimization algorithm for enhancing the extracted power under partial shadow condition. Energ. Rep. **7**, 109–124 (2021)
74. Fathy, A., Al-Dhaifallah, M., Rezk, H.: Recent coyote algorithm-based energy management strategy for enhancing fuel economy of hybrid FC/battery/SC system. IEEE Access **7**, 179409–179419 (2019)
75. Babu, N.R., Narrisetty, V., Saikia, L.C.: Maiden application of coyote optimizer algorithm with TIDN controller in AGC of a multi-area multi-source system. In: 2019 IEEE 16th India Council International Conference (INDICON), pp. 1–4. IEEE, Piscataway (2019)

𝕂 Semantics for Security Policy Enforcement on Android Applications with Practical Cases

Marwa Ziadia, Mohamed Mejri, and Jaouhar Fattahi

1 Introduction

Most Android users hinge on its adopted security solutions, such as the permission system to protect their data. Nevertheless, Android considers user privacy less important than users consider it. When installing an application, the user is faced with two choices: either they accept all the requested permissions, or they accept none; there is no middle ground. By refusing these permissions, Android will refuse to install the application. To make such a decision, users (and often even developers) usually do not have enough information to judge whether permission is really needed to perform a certain task, what is its true implication, its mala fide intentions (in case of spyware), and if it includes bugs/flaws. Indeed, these third-party applications have become a serious security threat. They gain access to numerous users' private data, demand more permissions than they really need. These threats to mobile security in general and user's privacy in particular, are becoming more and more imposing and raise many concerns. Their common objective is to enforce the security of downloaded applications and protect the user. Nevertheless, one of the main weaknesses of most initiatives is the absence of environments that allow formal specification, behavior verification, and policy enforcement. In this paper, we propose a formal framework for security policy enforcement on Android applications. In previous work [1], we presented an approach for security policy enforcement on reversed Android applications by rewriting. We emphasized the effectiveness of program rewriting techniques as they

M. Ziadia · M. Mejri · J. Fattahi (✉)
Department of Computer Science and Software Engineering, Laval University, Québec, QC, Canada
e-mail: marwa.ziadia.1@ulaval.ca; mohamed.mejri@ift.ulaval.ca; jaouhar.fattahi.1@ulaval.ca, https://www.ift.ulaval.ca/

M. D. Hina et al. (eds.), *Advances in Computational Intelligence and Communication*, EAI/Springer Innovations in Communication and Computing,
https://doi.org/10.1007/978-3-031-19523-5_8

derive from the synergy between the aspect-oriented paradigm and formal methods. These techniques combine the advantages of both: the rigor and unambiguity of using formal methods to specify the system, coupled with the ability of the aspect-oriented paradigm to automatically add the security aspect to that specification. In this paper, we endorse this idea and improve it with the use of a semantics engineering framework called $\mathbb{K}$ framework. This framework is armed with a set of tools. Some are useful during the language formalization process, others during the verification and syntactic analysis. Defining semantics with this environment allows to test it against a potentially large number of programs to gain confidence in its correctness. The included interpreter, semantics compiler, and debugger allow to avoid human errors and omissions that can slip through, and any possibility of escaping details while defining the semantics [2]. The approach proposed in [1] is theoretical and lacks such an environment to show its practical feasibility. In this paper, we study the automation of the entire process of policy enforcement using $\mathbb{K}$ framework. Our goal is to obtain automatically from an Android program and a security policy, a new version of the program that respects this policy. So all the necessary code will be added automatically in the appropriate positions in the program using an automatized tool, compelling it to abide by the policy. To reach this end, we implement the idea in $\mathbb{K}$ by defining syntax and semantics and we mechanize the whole enforcement process within the same environment.

The remainder of this paper is organized as follows. In Sect. 2, we recall some results achieved in previous work, and we introduce our approach through a figure that summarizes the adopted methodology. In Sect. 7, we discuss some work with comparable ideas. Section 3 presents the syntax and semantics of the logic modeling the security policy. Section 4 presents the formal language defined to specify an Android application. Section 5 details every theoretical step adopted for security policy enforcement and its equivalent in practice when using the $\mathbb{K}$ framework. In Sect. 6, we automatize the entire process by defining $\mathbb{K}$ semantics (syntax, configuration and rewrite rules). We wrap up this section with an example that shows the result of the enforcement technique automation. In Sect. 8, we conclude and point out the future avenues of our research.

2 Enforcement Process and Previous Work

In this section, we briefly recall essential details related to the approach adopted in [1] to enforce security policies on Android applications and we present our new methodology by using the $\mathbb{K}$ framework. In previous work, we presented three main steps for security policy enforcement on Android applications on which we have used formal methods. These steps consist of:

1. *Defining a dedicated logic for security policy.* We have used LTL (Linear Temporal Logic) for the policy specification.

2. *Defining formal language for Android applications*. We have presented 𝕂-Smali, a formal language adopted to specify the assembly-like code generated from reverse-engineering Android applications, Smali.
3. *Defining a formal technique for security policy enforcement*. We have developed a formal technique to automatically enforce an LTL formula on Smali using rewriting. For this step, we have proceeded by steps.
   (i) Transform the LTL formula obtained in 1 into a 𝕂-Smali program (a monitor) by adding synchronization or control actions. We have defined $\lceil - \rceil$ function for formula transformation;
   (ii) Modify the program obtained in 2 by including the complementary part of synchronization actions added to the formula. For this, we have defined $\lceil - \rceil\ mod\ -$ function;
   (iii) Apply the enforcement technique by running the two programs in parallel to force the synchronization. For this, we have defined an enforcement operator $\sqcap$, such as $P \sqcap_A \varphi = (\lceil P \rceil\ mod\ A) \parallel \lceil \varphi \rceil)$.

In this paper, we maintain the same procedure but we automatize it. More clearly, LTL formula transformation, program rewriting according to this formula, and finally the program rewriting according to the policy are all performed by automated tools provided by 𝕂. We also define syntax and semantics for LTL logic expressing the security policy. For the Smali program specification, we have already defined 𝕂-Smali using 𝕂. It is an executable semantics for Smali code. It was tested against sample programs (sequential and multi-threaded ones). Programs were also model-checked against LTL formulas using 𝕂 tools. For more details, we invite the reader to see [2]. 𝕂 is a rewriting-based definitional semantic framework for programming languages. It provides a complete methodology for their specification [2]. Different modules automatically derive from the same formal definition of the language. They include model-checking, symbolic execution, deductive program verification, etc. This environment also offers a plethora of tools, which are language independent but can be applied to any language having 𝕂 semantics. These tools include a parser, a state-space explorer, and a test-case generator. In this paper, we take advantage of 𝕂 tools to enforce the security of Android programs. For semantics design, 𝕂 provides a complete methodology. It is a strict guideline that consists of syntax, configuration, and semantics (or rewrite rules) definitions. Once done, definitions are saved in *k* extension files and compiled using a semantics compiler tool via the *kompile* command. Then, a *krun* command invokes an interpreter for program samples simulation. Many other options can be added to this command to generate other modules for formal verification. Figure 1 summarizes the main steps of our approach. The principal input is an Android application in the format *APK* (for "Android Package Kit"). This file is firstly converted to a readable format via a reverse-engineering tool. The result is a file *Smali* containing the code of the application accompanied by an *AndroidManifest* file in a readable representation. The second input is a security policy expressing a particular behavior. The 𝕂 framework takes the Smali program, manifest file, and the security policy. Then,

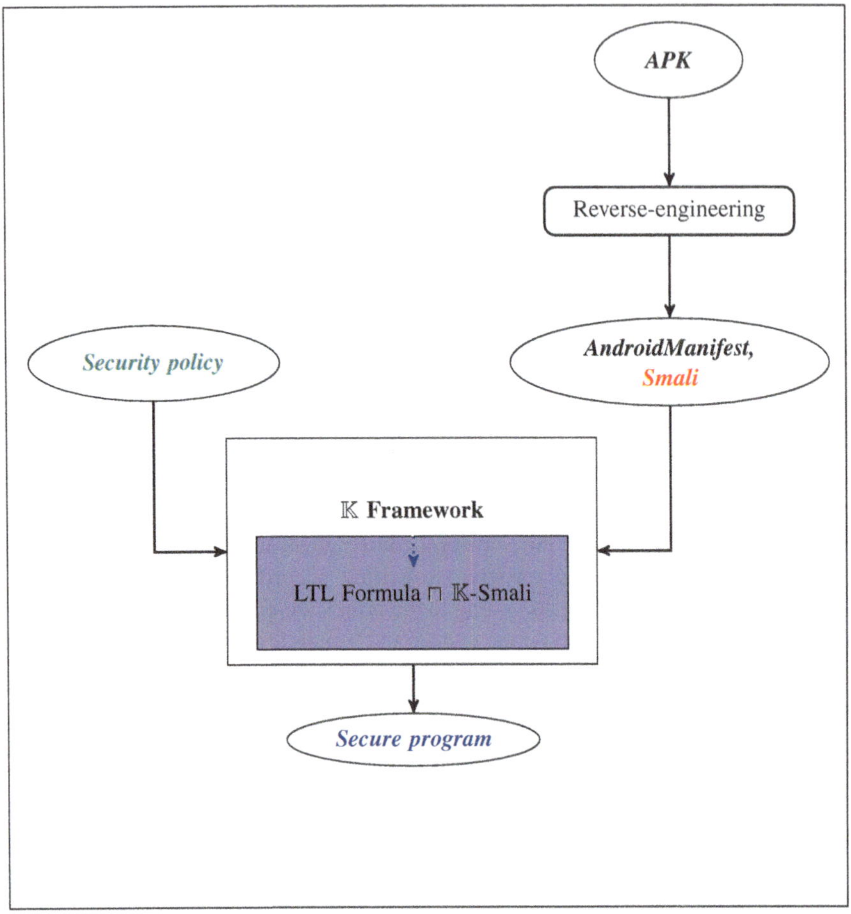

Fig. 1 Security policy enforcement process

generates an executable semantics, called 𝕂-Smali, for Smali with all the necessary information extracted from the *AndroidManifest* file, and an LTL logic expressing the security policy. In the same environment 𝕂, a new version of the program (denoted by *secure program*) that respects the policy is generated.

3 Security Policy Specification

Listing 1 presents the syntax of the logic adopted to specify a security policy. It is written with the BNF grammar. An LTL formula LTL can be the Boolean constants *True* or *False*. It can be also an empty action 1 or an atomic action $Action$. A formula can be the union or intersection of formulas. The *kleene* operator (*) is used for iteration.

Syntactic statements can be marked with attributes. They are specified in square brackets at the end of each definition and are intended to provide additional information to the parser. The attribute "left" annotates the definition of an LTL formula and used to mark the left associativity of the operators "+" and "." and "∗". The attribute "strict" specifies that LTL formulas must be evaluated before evaluating the construct itself, in other terms, it is non-deterministically strict in all enlisted formulas. Finally, the "bracket" attribute indicates that an LTL formula can be enclosed in brackets, which are generally used for grouping reasons.

```
module LTLFORMULA-SYNTAX
imports SMALI-SYNTAX
syntax   LTL::= "True"
              |"False"
              |"1"
              |Action
              |LTL "." LTL            [left, strict]
              |LTL "+" LTL            [left, strict]
              |LTL "*" LTL            [left, strict]
              |"(" LTL ")"            [bracket]
endmodule
```

4 Android Application Specification

In [2], we presented 𝕂-Smali, an executable semantics of Smali code using 𝕂-framework. Listing 2 represents its syntax. A Smali program consists of Smali and an *Androidmanifest* files. The latter is used to identify the application entry point (line 77). Smali files are a list of classes. Each class is composed of a header including all its information (possible comments, access flags, source and super class full names, and its fully qualified name). The class definition includes fields and methods as well. Methods are identified by their names, signature, and a body. The body consists of a list of directives and instructions. Directives are used to identify the number of registers used by the method (local and total registers). Instructions operate on the register's content. For more details on Smali instructions, registers manipulation, we refer the reader to [3].

Smali program configuration and semantics rules are also given in [2]. In this paper, we use directly 𝕂-Smali to apply LTL formulas.

```
1 module SMALI-SYNTAX
2  syntax Program     ::= SmaliFiles ManifestFile
3  syntax SmaliFiles  ::= List{SmaliFile," "}
4  syntax SmaliFile   ::= Class
5  syntax Class       ::= ClassHeader Fields Methods
6  syntax ClassHeader ::= Comments ".class" AccessFlags ClassName SuperClass
        SourceClass
7  syntax SuperClass  ::= Comments ".super" SuperClassName | Empty
8  syntax SourceClass ::= Comments ".source" String | Empty
9  syntax Comments ::= List{Comment," "}
```

```
syntax Comment  ::= r"\#.*"                                          [token]
syntax Fields ::= List{Field," "}
syntax Field ::= Comments ".field" AccessFlags FieldName ":" Type ValueOp
syntax ValueOp ::= Value | Empty
syntax Methods ::= List{Method," "}
syntax Method::=Comments ".method" AccessFlags MethodNameSign MethodBody ".
    endmethod"
syntax MethodNameSign ::= MethodName MethodSignature
syntax MethodSignature ::= MethodInTypes MethodRetType
syntax MethodInTypes ::= "(" Types ")" | "(" ")"
syntax MethodRetType ::= Type  | VoidType
syntax Type::= PrimitiveType | ObjectType | ArrayType
syntax PrimitiveType ::= "Z" | "B" | "C" | "D" | "F" | "I" | "J" | "S"
syntax VoidType  ::= "V"                    /* void type*/
syntax ObjectType ::= LName              /* Object reference*/
syntax ArrayType ::="["PrimitiveType|"["ObjectType|"[ArrayType"
syntax Value ::= Bool | Int | Float | String
syntax AccessFlags ::= List{AccessFlag," "}
syntax AccessFlag::= "public"|"private"|"protected" |"final"|"abstract"|"
    static"
syntax ClassName ::= LName
syntax SuperClassName::= LName
syntax MethodName ::= Name | "constructor" "<init>"
syntax FieldName::=Name
syntax Name  ::= Id
syntax LName ::= r"L[_a-zA-Z0-9]*[_a-zA-Z0-9]*;"                     [token]
syntax MethodRef  ::= ClassName "->" MethodNameSignature
syntax FieldRef   ::= ClassName "->" FieldName
syntax Parameters ::= List{Parameter,","}
syntax Parameter  ::= RegName
syntax MethodBody ::= List{Statement," "}
syntax Statement  ::= Directive | Instruction
syntax Directive  ::= ".locals" Int | ".registers" Int
syntax Instruction::= "goto" ":" Label
                   |":" Label
                   |"nop"
                   |"sparse-switch" RegName "," ":" Switchtab
                   |"const" RegName "," Val
                   |"const-string" RegName "," String
                   |"move" RegName "," RegName
                   |"new-instance" RegName "," ClassName
                   |"new-array" RegName "," RegName "," ArrayType
                   |Sget RegName "," FieldRef
                   |Sput RegName "," FieldReference
                   |"iget" RegName "," RegName "," FieldRef
                   |"iput" RegName "," RegName "," FieldRef
                   |"aget" RegName "," RegName "," RegName
                   |"aput" RegName "," RegName "," RegName
                   |"if-eq" RegName "," RegName "," ":" Label
                   |"if-lt" RegName "," RegName "," ":" Label
                   |BinOp RegName "," RegName "," RegName            [left]
                   |UnOp RegName "," RegName
                   |"invoke-static" "{"Parameters"}" "," MethodRef
                   |"invoke-virtual" "{"Parameters"}""," MethodRef
                   |"move-result" RegName
                   |"retrun-void"
                   |"return" RegName
                   |"monitor-enter" RegName
                   |"monitor-exit" RegName
syntax Sput  ::= "sput" | "sput-object"
syntax Sget  ::= "sget" | "sget-object"
syntax Binop ::= "add" | "sub" | "mul" | "div" |...
syntax Unop  ::= "neg" | "not" | "int-to-long" |...
syntax Val  ::= Int
syntax Switchtab ::= ".sparse-switch" Tablecases ".end sparse-switch"
syntax Tablecases ::= List {Tablecase," "}
syntax Tablecase ::= Value "->" ":" Label
```

```
  syntax StringId , Label ::= Id
  syntax Empty ::= " "
  syntax ManifestFile ::=".manifest" MethodRef
endmodule
```

Listing 2 𝕂-Smali: syntaxe

5 𝕂 Semantics for Security Policy Enforcement on Smali

In this section, we show how we use 𝕂 framework to enforce security policy on Android applications. As already mentioned, we pursue the same methodology that was adopted in [1].

1. $\lceil P \rceil$ *mod* A: *Rewrite the program P by adding some control actions according to the security policy* φ *and to actions in* A.
   This step is implemented in 𝕂 by adding synchronization methods called *send* and *receive*. These methods create channels between the program and the policy so they can communicate and synchronize. Methods *receive* from a *start channel* and *send* on an *end channel* will be added before and after each action considered by the policy, respectively. Listing 3 presents the code of the *send* and *receive* methods implemented in 𝕂. They take as parameter a channel (a String object). Both methods allow the synchronization between threads through *monitor-enter* and *monitor-exit* Smali instructions (lines 8, 16, 24, 27). These two instructions ensure the mutually exclusive access to shared objects by different threads [3] and in our case between two threads that represent the program and the policy. This principle has been used to ensure that only one thread (either the program or the policy) owns the monitor for a shared object (i.e., the channel). This way, only one thread that acquires the monitor can send or receive from a shared channel. Others are blocked until the release of the object's (channel) monitor.
2. $[\![\varphi]\!]$: *Transform the security policy* φ *into a monitor or a* 𝕂*-Smali program.*
   This step is implemented in 𝕂 by creating a new thread that represents the security policy, including the complementary part of synchronization methods added to P (i.e., *send* on a *start channel* followed by a *receive* from an *end channel*). The purpose behind adding these methods to the program and the security policy, in this order, is to create a kind of communication between the two via a channel. In this communication, full control is given to the policy, it is the policy that can start communication by a *send* on a start channel and it is the policy that can close it once receiving from the end channel. The supervised program, from its side, waits for the authorization of the policy. It starts its execution as soon as it receives a signal from the policy on the start channel (since it always starts with a *receive*). This way, the program is always blocked waiting for a signal from the policy, and the policy has total control and can block and unblock the execution of specific actions, as wanted.

3. $\mathbf{P} \sqcap_{\mathbf{A}} \varphi = (\lceil \mathbf{P} \rceil \ \mathbf{mod}\ \mathbf{A}) \parallel \lceil \varphi \rceil$: *Running the two programs in parallel to synchronize.*

   In $\mathbb{K}$, this step is carried out by launching threads (i.e., the program threads and the security policy).

```
.class public Lp/PhiChannel;
.field static a: LJava/lang/String;

.method public static send(Ljava/lang/String;)V
 .locals 3
 sget-object v0, Lp/c3;->x
 :Wait
 monitor-enter v0
 iget v1, v0, Lp/PhiChannel;->a
 const-string v2, "undefined"
 if-eq v1,v2, :label1
 monitor-exit v0
 goto :Wait
 :label1
 iput p0, v0, Lp/PhiChannel;->a
 monitor-exit v0
 retrun-void
.end method

.method public static receive(Ljava/lang/String;)V
 .locals 3
 sget-object v0, Lp/c3;->x
 :Wait
 monitor-enter v0
 iget v1, v0, Lp/PhiChannel;->a
 if-eq v1,p0, :label1
 monitor-exit v0
 goto :Wait
 :label1
 const-string v2, "undefined"
 iput v2, v0, Lp/PhiChannel;->a
 monitor-exit v0
 return-void
.end method
```

Listing 3 Synchronization methods (*send, receive*)

To further clarify the idea, we illustrate through an example how to enforce a security policy on a given program using $\mathbb{K}$ framework. The whole necessary code added to enforce the policy on the program is given in this example. Once all steps are made, we use the $\mathbb{K}$ framework interpreter to run the enforced program. We show how it behaves as required by the policy as well.

The program P in Listing 4 consists of three classes $Lp/c1$;, $Lp/c2$; and $Lp/c3$. The *.manifest* (line 35) indicates the program entry point (i.e., the method $m1$ from the class c3). The first class includes a method *start*. The method body consists of displaying the strings "a" and then "b" by calling the method *println*. The second class includes another method *start* that invokes the *println* method on strings "c" and then "d." The class $Lp/c3$; is a sub-class of the class $Lp/c1$;. Two threads are instantiated from $Lp/c1$; and $Lp/c2$; (lines 28 and 30) and launched (lines 29 and 31). Once started and executed in parallel, they display string constants in the following order "a" "b" "c" "d."

Given a security policy $\varphi = $ "$a.c.b.d$" that requires the display of strings in the following order "a" "c" "b" "d " and the set of actions to be controlled A={constant-string, println}.

To enforce φ on P in $\mathbb{K}$, we proceed as follows:

1. $\lceil P \rceil$ *mod* A
   Modify the program P by inserting the synchronization actions (*send* and *receive*) in specific points that are determined by the security policy and specified in A. These points correspond to the display of constant strings "a," "c," "b," and "d" with the instruction *invoke-virtual* of the method *println* (lines 6, 8, 17, 19 in Listing 4). So, each of these actions in P will be preceded and followed by calls to the synchronization methods *receive* and *send* from the class *Lp/PhiChannel;* (see listing 3), respectively. This step is marked with blue in Listing 5. The method $receive$ will be invoked to receive from a start channel a_s, b_s, c_s, or d_s, whereas the invoked $send$ method takes as parameter an end channel a_e, b_e, c_e, or d_e.
2. $\lceil \varphi \rceil$
   Transform the security policy into a monitor that could synchronize with P. The policy is included in a new class *Lp/Phi;* inserted into the program code. This step is marked with green in Listing 5. The class includes the complementary part of synchronization actions (*send* followed by a *receive*) in the order required by the policy ("a," "c," "b," "d").
3. $(\lceil P \rceil$ *mod* $A) \parallel \lceil \varphi \rceil$
   To enforce φ on P, we run the modified program and the policy in parallel. A new thread instantiated from the class Lp/Phi; (including the security policy) will be launched to execute in parallel with the program's threads. The code presented with red color in Listing 5 represents instructions added to run the controller thread in parallel with the transformed program. In line 75, the new thread is instantiated from the class Lp/Phi; and then launched in line 76.

Creating channels from which the policy can control the program and block some of its "risky" actions was inspired from the *BlockingQueue* [4] used in Java for concurrent execution. It is a queue, on which multiple threads can be inserting and taking elements concurrently, without any concurrency issues arising. The blocking aspect comes when a thread tries to enqueue or dequeue an element and there are none left in the queue, then it is blocked until there is an element to take. In the same stream of thought, the start and the end channel are equivalent to the blocking queue. The program cannot advance until it has received confirmation on a start channel from the policy. Similarly, the policy cannot process until it has received the end mark from the program on an end channel.

When we run the enforced program P' (Lisitng5) using the $Krun$ command in $\mathbb{K}$, we notice the synchronization between the security policy included in the newly added thread and the modified program as expected. We see that the new program respects the enforced policy, i.e., the display of "a.c.b.d" as stated by the policy. The result is illustrated in Listing 6 from traces generated when running P'. For the sake of simplicity and to make the produced result easy to comprehend, we have simplified and kept only traces affected by the policy. Traces shown in lines 9, 18, 27, and 35 where strings "a," "c," "b," "d" are, respectively, highlighted and displayed in the required order.

```
.class public Lp/c1;
.method public static start()V
 .locals 2
  sget-object v0, Ljava/lang/System;->out:Ljava/io/PrintStream;
  const-string v1, "a"
  invoke-virtual {v0,v1},Ljava/io/PrintStream;->println(Ljava/lang/String;)V
  const-string v1, "b"
  invoke-virtual {v0,v1},Ljava/io/PrintStream;->println(Ljava/lang/String;)V
  return-void
.end method

.class public Lp/c2;
.method public static start()V
 .locals 2
  sget-object v0,Ljava/lang/System;->out:Ljava/io/PrintStream;
 const-string v1, "c"
 invoke-virtual {v0,v1},Ljava/io/PrintStream;->println(Ljava/lang/String;)V
 const-string v1, "d"
 invoke-virtual {v0,v1},Ljava/io/PrintStream;->println(Ljava/lang/String;)V
 return-void
.end method

.class public Lp/c3;
.super Lp/c1;
.field public x: Ljava/lang/Object;
.method public static m1()V
  .locals 2
  new-instance v0, Lp/c1;
  invoke-virtual {v0}, Ljava/lang/Thread;->start()V
  new-instance v1, Lp/c2;
  invoke-virtual {v1}, Ljava/lang/Thread;->start()V
  return-void
.end method

.manifest Lp/c3;->m1()V
```

Listing 4 Original program *P* before policy enforcement

```
.class public Lp/c1;
.method public static start()V
.locals 2
sget-object v0, Ljava/lang/System;->out:Ljava/io/PrintStream;
const-string v1, "a_s"
invoke-static {v1}, Lp/PhiChannel;->receive(Ljava/lang/String;)V
const-string v1, "a"
invoke-virtual {v0, v1}, Ljava/io/PrintStream;->println(Ljava/lang/String;)
const-string v1, "a_e"
invoke-static {v1}, Lp/PhiChannel;->send(Ljava/lang/String;)V

const-string v1, "b_s"
invoke-static {v1}, Lp/PhiChannel;->receive(Ljava/lang/String;)V
const-string v1, "b"
invoke-virtual {v0, v1}, Ljava/io/PrintStream;->println(Ljava/lang/String;)
const-string v1, "b_e"
invoke-static {v1}, Lp/PhiChannel;->send(Ljava/lang/String;)V
return-void
.end method

.class public Lp/c2;
```

```
 .method public static start()V
 .locals 3
 const-string v1, "c_d"
 invoke-static {v1}, Lp/PhiChannel;->receive(Ljava/lang/String;)V
 const-string v1, "c"
 invoke-virtual {v0, v1}, Ljava/io/PrintStream;->println(Ljava/lang/String;)
 const-string v1, "c_e"
 invoke-static {v1}, Lp/PhiChannel;->send(Ljava/lang/String;)V

 const-string v1, "d_s"
 invoke-static {v1}, Lp/PhiChannel;->receive(Ljava/lang/String;)V
 const-string v1, "d"
 invoke-virtual {v0, v1}, Ljava/io/PrintStream;->println(Ljava/lang/String;)
 const-string v1, "d_e"
 invoke-static {v1},Lp/PhiChannel;->send(Ljava/lang/String;)V
  return-void
.end method

.class public Lp/Phi;
.method public static start()V
 .locals 2
 const-string v0, "a_s"
 invoke-static {v0}, Lp/PhiChannel;->send(Ljava/lang/String;)V
 const-string v0, "a_e"
 invoke-static {v0}, Lp/PhiChannel;->receive(Ljava/lang/String;)V
 const-string v0, "c_s"
 invoke-static {v0}, Lp/PhiChannel;->send(Ljava/lang/String;)V
 const-string v0, "c_e"
 invoke-static {v0}, Lp/PhiChannel;->receive(Ljava/lang/String;)V
 const-string v0, "b_s"
 invoke-static {v0}, Lp/PhiChannel;->send(Ljava/lang/String;)V
 const-string v0, "b_e"
 invoke-static {v0}, Lp/PhiChannel;->receive(Ljava/lang/String;)V
 const-string v0, "d_s"
 invoke-static {v0}, Lp/PhiChannel;->send(Ljava/lang/String;)V
 const-string v0, "d_e"
 invoke-static {v0}, Lp/PhiChannel;->receive(Ljava/lang/String;)V
 return-void
.end method

.class public Lp/c3;
.super Lp/c1;
.field public x: Ljava/lang/Object;
.method public static ml()V
  .locals 3
  new-instance v0, Lp/PhiChannel;
  const-string v1, "undefined"
  iput v1, v0, Lp/PhiChannel;->a
  sput-object v0, Lp/c3;->x
  new-instance v0, Lp/c1;
  invoke-virtual {v0}, Ljava/lang/Thread;->start()V
  new-instance v1, Lp/c2;
  invoke-virtual {v1}, Ljava/lang/Thread;->start()V
  new-instance v2, Lp/Phi;
  invoke-virtual {v2},Ljava/lang/Thread;->start()V
  return-void
.end method

.manifest Lp/c3;->ml()V
```

Listing 5 Enforced program P'

```
<trace>
 \rconstString { Lp/c3; -> ml ( )V } { v1 } {"undefined"}
 \rconstString { Lp/c1; -> start ( )V } { v1 } { "as" }
 \rconstString { Lp/c2; -> start ( )V } { v2 } { "cs" }
 \rconstString { Lp/Phi; -> start ( )V } { v0 } { "as" }
 \rconstString { Lp/PhiChannel; ->send( Ljava/lang/String;)V } {v2} {"
     undefined"}
 \rconstString { Lp/Phi; -> start ( ) V } { v0 } { "ae" }
 \rconstString { Lp/PhiChannel; ->receive(Ljava/lang/String;)V }{v2}{"
     undefined"}
 \rconstString { Lp/c1; -> start ( )V } { v1 } {"a" }
 \rconstString { Lp/c1; -> start ( )V } { v1 } { "ae" }
 \rconstString { Lp/PhiChannel; ->send (Ljava/lang/String;)V} {v2}{"
     undefined"}
 \rconstString { Lp/c1; -> start ( )V } { v1 } { "bs" }
 \rconstString { Lp/PhiChannel;->receive(Ljava/lang/String;)V } {v2} {"
     undefined"}
 \rconstString { Lp/Phi; -> start ( )V } { v0 } { "cs" }
 \rconstString { Lp/PhiChannel; ->send (Ljava/lang/String;)V} {v2} {"
     undefined"}
 \rconstString { Lp/Phi; -> start ( )V } { v0 } { "ce" }
 \rconstString { Lp/PhiChannel;->receive(Ljava/lang/String;)V} {v2} {"
     undefined"}
 \rconstString { Lp/c2; -> start ( )V } { v2 } {"c" }
 \rconstString { Lp/c2; -> start ( )V } { v2 } { "ce" }
 \rconstString { Lp/PhiChannel; ->send(Ljava/lang/String;)V} {v2} {"
     undefined"}
 \rconstString { Lp/PhiChannel;->receive(Ljava/lang/String;)V }{v2}{"
     undefined"}
 \rconstString { Lp/c2; -> start ( )V } { v2 } { "ds" }
 \rconstString { Lp/Phi; -> start ( )V } { v0 } { "bs" }
 \rconstString { Lp/PhiChannel;->send(Ljava/lang/String;)V} {v2}{"undefined"}
 \rconstString { Lp/PhiChannel;->receive( Ljava/lang/String;)V}{v2} {"
     undefined"}
 \rconstString { Lp/Phi; -> start ( )V } { v0 } { "be"}
 \rconstString { Lp/c1; -> start ( )V } { v1 } {"b" }
 \rconstString { Lp/c1; -> start ( )V } { v1 } { "be" }
 \rconstString { Lp/PhiChannel;-> send(Ljava/lang/String; )V} {v2} {"
     undefined"}
 \rconstString { Lp/PhiChannel;->receive(Ljava/lang/String;)V }{v2} {"
     undefined"}
 \rconstString { Lp/Phi; -> start ( )V } { v0 } { "ds" }
 \rconstString { Lp/PhiChannel;-> send(Ljava/lang/String;)V}{v2}{"undefined"}
 \rconstString { Lp/PhiChannel; ->receive(Ljava/lang/String;)V }{v2}{"
     undefined"}
 \rconstString { Lp/Phi; -> start ( )V } { v0 } { "de" }
 \rconstString { Lp/c2; -> start ( )V } { v2 } {"d"}
 \rconstString { Lp/c2; -> start ( )V } { v2 } { "de" }
 \rconstString { Lp/PhiChannel;->send (Ljava/lang/String; )V} {v2}{"
     undefined"}
 \rconstString { Lp/PhiChannel; ->receive(Ljava/lang/String;)V}{v2}{"
     undefined"}
</trace>
```

Listing 6 Traces generated using $\mathbb{K}$ Framework interpreter on P'

6 Automation Using $\mathbb{K}$ Semantics

In the previous example, we showed how to enforce a security policy on a given program in practice. The code in this example was injected manually in the original program following the mentioned three steps. In this section, we study the automation of the entire process of policy enforcement using the same environment 𝕂.

The objective is to obtain automatically, from a program and a formula, a new program that behaves as stated by the policy. So, all the necessary code for the enforcement will be added automatically in the right positions in the program. To this end, we implemented the idea in 𝕂 by defining syntax, configuration, and semantics rules.

6.1 Syntax

Listing 7 represents the module ENFORCMEMENT-SYNTAX. It represents the 𝕂 syntax definition for formula enforcement on 𝕂-Smali program. This module calls modules SMALI-SYNTAX (Listing 2) and LTLFORMULA-SYNTAX (Listing 1) imported for the program and the formula definitions. The enforcement syntax consists of 𝕂-Smali program "*Program*" "mod" a list of actions to be controlled "[*Actions*]" enforced "⊓" by an LTL formula "LTL".

```
module ENFORCEMENT–SYNTAX
imports SMALI–SYNTAX
imports LTLFORMULA–SYNTAX
syntax Actions ::= List "{" Action ," " "}"
syntax Action  ::= Instructions
syntax LTL     ::= LtlFormula | Action
syntax Enforcement ::= Program "mod"  "["Actions"]" "<"⊓">" LTL
end module
```

Listing 7 𝕂 syntax for formula enforcement on 𝕂-Smali

6.2 Configuration

In 𝕂, the state of a running program is represented by a configuration. Configurations are structured as labeled and potentially nested cells.

Figure 2 represents the enforcement configuration of a program by a formula. It consists of a top cell labeled ⊤, containing two sub-cells: the cell k contains the running program *PGM* of type k which can also be the formula before its transformation into a monitor. The second cell labeled *EnforcedProgram* contains the enforced program or the transformed formula. It represents the result after the modifications made to the program or formula in the k cell in order to implement the enforcement. This configuration can be applied to any program having 𝕂 semantics enforced by a security policy. The first cell represents the unsafe program of type K, while the second cell contains the enforced program.

Fig. 2 Formula enforcement on $\mathbb{K}$-Smali program configuration

6.3 Semantics

Defining semantics for the language constructs consists of a set of $\mathbb{K}$ rewrite rules that describe the evolution of a running configuration as a transition from an initial configuration holding the original program to a new one maintaining the result. The rule has the form rule $C \Rightarrow C'$. Intuitively, if C is a fragment of the current state of the configuration, then the rule can apply and it will be replaced by the new configuration fragment C'. We define all $\mathbb{K}$ rewrite rules needed for LTL formula transformation, program rewriting according to the security policy, and the synchronization between them. $\mathbb{K}$ rules used for policy enforcement are depicted in Fig. 3. The rule $R_{LTL-to-Smali}$ represents the transformation of LTL formula transformed into a $\mathbb{K}$-Smali program. The formula 1 is transformed into its equivalent instruction in $\mathbb{K}$-Smali, *nop*. The rule $R_{LTL-to-Smali}$ transforms a *get* action into a new formula that can synchronize once enforced on the program, by calling *send* and *receive* methods. $\mathbb{K}$ enforcement rules rewrite the introduced program automatically by adding appropriate synchronization actions in their corresponding locations. These actions are held in a class injected into the program called each time through its fully qualified name. From an original program consisting of a *goto* instruction and an LTL formula consisting of two consecutive instructions *goto* and *iput*, the rule $R_{Enforcement}$, rewrites the entry code by a new sequence of instructions in which *receive* and *send* are added before and after *goto Lab* instruction, respectively.

6.4 Example

As we have just explained, we define an intersection operator ⊓ that takes as inputs a program, a set of actions to be controlled, an LTL formula, and generates a new program that satisfies the formula. Let us take an example to see how the policy enforcement on the $\mathbb{K}$-Smali was automatized. Listing 8 presents an example of inputs. It consists of a $\mathbb{K}$−Smali program composed of a class $Lp/cl2;$. Then, in line 17, we found the enforcement operator "⊓" followed by the LTL formula. It consists of a sequence of actions (or instructions), including composition "." and iterative "*" operators. The modulo operator "mod" identifies actions to be

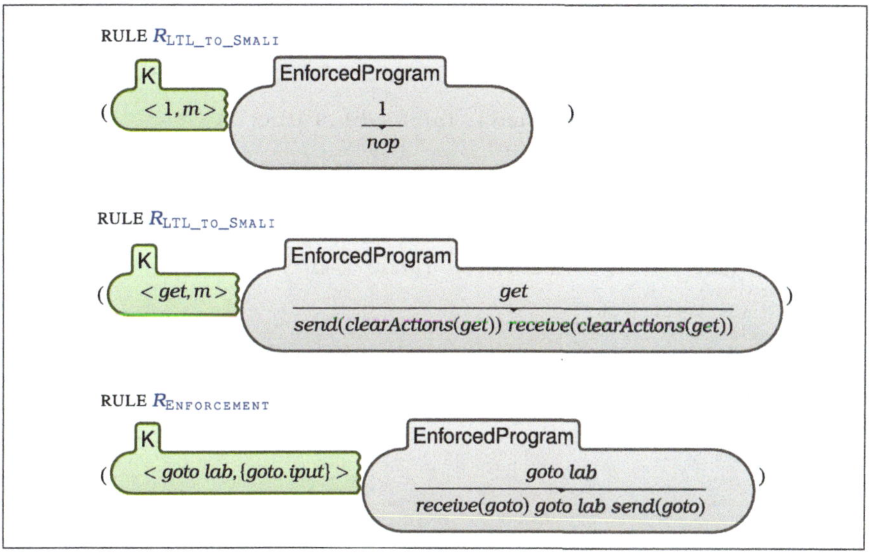

Fig. 3 𝕂 rewrite rules for formula enforcement

controlled. Besides the actions specified by the formula, they will be controlled by adding synchronization actions before and after each one.

```
.class public Lp/cl2;
.super Lp/cl;
.field public x: I
.field public y: C
.method public static ml()V
 .locals 3
 const vl, 30
 goto L1
 :L1
 invoke-static {v0, vl}, Lp/cl;->m2(I C)C
 sput v0, Lp/c2;->x
 new-instance v2, Lp/cl;
 invoke-virtual {v2,vl}, Lp/cl;->m2(I C)C
 iput vl, v2, Lp/cl;->b
 return-void
 .end method

 mod {new-instance goto invoke-virtual invoke-static}

 <□>

(new-instance.invoke-static.invoke-virtual)*
    (new-instance.invoke-static)
```

Listing 8 Enforcement inputs

```
<T>
 <k>•</k>
 <EnforcedProgram>
 .locals 3
 const vl, 30
```

```
6 receive(goto)
7 goto L1
8 send(goto)
9 :L1
10 receive(invoke-static)
11 invoke-static {v0, v1 .Parameters}, Lp/c1;->m2( I C )C
12 send(invoke-static)
13 sput v0,Lp/c2;->x
14 receive(new-instance)
15 new-instance v2, Lp/c1;
16 send(new-instance)
17 receive(invoke-virtual)
18 invoke-virtual {v2, v1, .Parameters}, Lp/c1; ->m2( I C )C
19 send(invoke-virtual)
20 iput v1, v2, Lp/c1; ->b
21 return-void
22 .end method
23
24 .Comments .method public Phi ( ) V
25 :Lab0
26 send(new-instance )
27 receive(new-instance)
28 send(invoke-static)
29 receive(invoke-static )
30 send(invoke-virtual )
31 receive(invoke-virtual )
32 goto Lab0
33 .end method
34
35 new-instance v2, Lp/Phi;
36 invoke-virtual {v2},Ljava/lang/Thread;->start()V
37 </EnforcedProgram>
38 </T>
```

Listing 9 Program rewriting

After introducing the formula and the program, we use the 𝕂 interpreter invoked through the *krun* command. Listing 9 represents the configuration resulting to this command. In the *enforcedProgram* cell (from line 3 to line 42), we see the new program after integrating the LTL formula. (1) Each action stated by the formula is preceded and followed by the corresponding synchronization actions (send and receive) in green. (2) In blue the formula transformation into 𝕂-Smali code with the corresponding control action *send* and *receive*. Notice the new lab *Lab*0 created (line 25) and the unconditional jump to this lab (with *goto* line 32). It is a new label generated by semantics rules to handle the iterative behavior of the formula *(new-instance.invoke-static.invoke-virtual)**. (3) In (red), a new thread holding the formula code is instantiated and started in lines 40 and 41. All these modifications to the program compelling it to abide by the policy were made by 𝕂.

7 Related Work and Discussion

Android relies on different security solutions to manage its applications safely, such as the permission system and the Security Enhancements for Android (SEAndroid) adopted to enforce a set of policy rules, including mandatory access control (MAC) and Discretionary Access Control (DAC) policies, on Android devices. Despite this, Android continues to report vulnerabilities related to these solutions. In [5], Yu et al. analyze the SEAndroid. They affirm that its original security policy rules do not pose any problem. The risk, however, comes from customized and added ones by mobile device manufacturers. Authors use SEPAL, a universal tool to automatically retrieve and investigate the customized policy rules. Results identify 7,111 unregulated policy rules with a low false-positive rate from 595,236 customized rules and reveal that the policy customization problem is worsening in newer Android versions. In [6], Lee al. propose PolyScope, a tool for the analysis of Android file system access control. It identifies Android access control policies and relevant system configurations to determine the attack operations available to adversaries. PolyScope was applied to triage three Google and five OEM (Original Equipment Manufacturer) Android releases. Results show a significantly greater number of attack operations for OEM's Android releases. By default, Android places installed third-party applications in one single domain called "untrusted_app". All applications in this domain are controlled by one predefined MAC policy. As result, when it authorizes an application to access a specific resource, all other applications in the domain are also allowed. As consequence, an adversary third-party Android application can access unauthorized resources on the device. Similarly, untrusted applications can access data from sensitive third-party applications, which violates the principle of least privilege. In [7] Pande et al. implement a new exclusive domain in which they place only sensitive applications from the untrusted domain. Then, they propose a more fine-grained and stricter MAC policy and enforce it on these applications. In [8], they focus on information security threats that target enterprises, particularly government organizations, financial institutions, and research institutes, where data is more sensitive and valuable. The risk mainly stems from the use of Android-based devices by employees in offices. To solve this problem, they propose solutions that preserve the use of Android smartphones, but attempt to enhance their security within a virtual geographic periphery, called a geofence. Inside this area, Android applications' access to various smartphone resources is significantly restricted. This is achieved through geofence-specified security policy enforcement. Some approaches chose to integrate security policies into the operating systems (OSs). In [9], Schlegel affirms that applications need individual security policies with application-specific semantics. The author claims that current approaches for application-level security policy enforcement suffer from the large and complex trusted computing bases issue where correctness and tamper proofness are difficult to achieve. To address this problem, he proposes AppSPEAR, a security architecture tailored to application-level security policy enforcement, on which he focuses on access control policy. All these attempts to improve the security solutions adopted by the Android system cannot be proven correct since they are not based on its formal specification. Formal methods fill these gaps and guarantee that specification covers as much as possible details, especially when it is generated by a semantics definitional framework like 𝕂. Costa et al. [10] propose PolEnA, an extension of their Android framework security presented in [11] for enforcing permissions policies in Android. PolEnA defines a formal language for security policies. To ensure compatibility with the tool, the authors propose a rewriting framework. An APK is firstly decompiled, then the framework rewrites its Java bytecode. Finally, the application is repackaged and signed. In [12], Talegaon et al. present a formal specification of the Android access control system. Defined language focuses on permissions and Uri permissions and includes two parts, User-Initiated Operations (UIOs) and Application-Initiated Operations (AIOs). During the formalization process, authors discover some peculiar behaviors in Android's access control system but do not present any alternative to enforce the security of such behaviors. In [13], Betarte et al. present as well a formal specification of Android's permission model. However, the resulted formal language was used to enforce permission-based access control policies. Authors use the Coq proof assistant [14] to verify security-related proprieties, such as eavesdropping and intent spoofing. Although the above approaches are based on a formal foundation of the Android system, they are limited to one part, namely the permission system. Therefore, the enforced security policies are restricted to this part and cannot address other vulnerabilities related to other aspects in Android. However, the formalization of Smali and AndroidManifest files allows applying various security policies, since the behavior of the application is all in these files.

8 Conclusion

In this paper, we have presented an approach for security policy enforcement on Android applications. We have formalized and implemented it using the $\mathbb{K}$ framework. This environment was used as a proof of concept of the approach that takes an Android application and a security policy as inputs, applies a sequence of modifications and adjustments, and generates a new version of the program that achieves the policy goals. From the introduced formula and in an elegant way, the semantics allows to identify automatically the relevant points in the code in which security controls should be inserted. The proposed solution for security enforcement on Android applications will undoubtedly safeguard the user's private life from cyber-attacks and keep serenity when downloading or purchasing new applications. A complete proof of results achieved in this work is actively underway.

Acknowledgments This research was funded by Natural Sciences and Engineering Research Council of Canada (NSERC) grant number RGPIN/05948-2016. The authors would like to thank Mrs. Samia Loukil for her proofreading of the manuscript.

References

1. Ziadia, M., Mejri, M., Fattahi, J.: Formal and automatic security policy enforcement on android applications by rewriting. In: Fujita, H., Pérez-Meana, H. (eds.) New Trends in Intelligent Software Methodologies, Tools and Techniques – Proceedings of the 20th International

Conference on New Trends in Intelligent Software Methodologies, Tools and Techniques, SoMeT 202, Cancun, 21–23 September 2021, pp. 337:85–98 (2021)
2. Ziadia, M., Mejri, M., Fattahi, J.: k-smali: an executable semantics for program verification of reversed android applications. In: Manuscript Submitted to the 14th International Symposium on Foundations and Practice of Security, FSP'2021 (2021)
3. Ziadia, M., Fattahi, J., Mejri, M., Pricop, E.: Smali+: an operational semantics for low-level code generated from reverse engineering android applications. Information **11**(3) (2020)
4. Java™Platform Standard Ed. 7. Interface blockingQueue<e>. https://docs.oracle.com/javase/7/docs/api/java/util/concurrent/BlockingQueue.html. Accessed 05 Jan 2021
5. Yu, D., Yang, G., Meng, G., Gong, X., Zhang, X., Xiang, X., Wang, X., Jiang, Y., Chen, K., Zou, W., Lee, W., Shi, W.: SEPAL: towards a large-scale analysis of SEAndroid policy customization. In: Leskovec, J., Grobelnik, M., Najork, M., Tang, J., Zia, L. (eds.) WWW '21: The Web Conference 2021, Virtual Event/Ljubljana, April 19–23, 2021, pp. 2733–2744. ACM/IW3C2, New York (2021)
6. Lee, Y.T., Enck, W., Chen, H., Vijayakumar, H., Li, N., Wang, D., Qian, Z., Petracca, G., Jaeger, T.: PolyScope: multi-policy access control analysis to triage android systems. CoRR, abs/2008.03593 (2020)
7. Pande, P., Mallaiah, K., Gandhi, R.K., Medatiya, A.K., Srinivasachary, S.: Fine grained confinement of untrusted third-party applications in android. In: 2021 International Conference on Computing, Communication, and Intelligent Systems (ICCCIS), pp. 372–376 (2021)
8. Pande, P., Medatiya, A.K., Mallaiah, K., Gandhi, R.K., Srinivasachary, S.: Mandatory enforcement of geofenced security in android. In: 2021 International Conference on Artificial Intelligence and Smart Systems (ICAIS), pp. 1031–1035 (2021)
9. Schlegel, M.: Trusted enforcement of application-specific security policies. CoRR, abs/2105.01970 (2021)
10. Costa, G., Sinigaglia, F., Carbone, R.: Polena: enforcing fine-grained permission policies in android. In: Tonetta, S., Schoitsch, E., Bitsch, F. (eds.) Proceedings of the Computer Safety, Reliability, and Security – SAFECOMP 2017 Workshops, ASSURE, DECSoS, SASSUR, TELERISE, and TIPS, Trento, September 12, 2017, vol. 10489. Lecture Notes in Computer Science, pp. 407–414. Springer, Berlin (2017)
11. Armando, A., Carbone, R., Costa, G., Merlo, A.: Android permissions unleashed. In: Fournet, C., Hicks, M.W., Viganò, L. (eds.) IEEE 28th Computer Security Foundations Symposium, CSF 2015, Verona, 13–17 July, 2015, pp. 320–333. IEEE Computer Society, Washington (2015)
12. Talegaon, S., Krishnan, R.: A formal specification of access control in android. In: Sahay, S.K., Goel, N., Patil, V., Jadliwala, M., (eds.) Proceedings of the Secure Knowledge Management in Artificial Intelligence Era – 8th International Conference, SKM 2019, Goa, December 21–22, 2019, vol. 1186. Communications in Computer and Information Science, pp. 101–125. Springer, Berlin (2019)
13. Betarte, G., Campo, J.D., Cristiá, M., Gorostiaga, F., Luna, C., Sanz, C.: Towards formal model-based analysis and testing of android's security mechanisms. In: Monteverde, H., Santos, R. (eds.) 2017 XLIII Latin American Computer Conference, CLEI 2017, Córdoba, September 4–8, 2017, pp. 1–10. IEEE, Piscataway (2017)
14. Bertot, Y., Castéran, P.: Interactive Theorem Proving and Program Development – Coq'Art: The Calculus of Inductive Constructions. Texts in Theoretical Computer Science. An EATCS Series. Springer, Berlin (2004)

Author Index

M. D. Hina et al. (eds.), *Advances in Computational Intelligence and Communication*, EAI/Springer Innovations in Communication and Computing,
https://doi.org/10.1007/978-3-031-19523-5

Subject Index

M. D. Hina et al. (eds.), *Advances in Computational Intelligence and Communication*, EAI/Springer Innovations in Communication and Computing,
https://doi.org/10.1007/978-3-031-19523-5

P

R

S

T

V

The manufacturer's authorised representative in the EU is Springer Nature Customer Service Centre GmbH, Europaplatz 3, 69115 Heidelberg, Germany. If you have any concerns regarding our products, please contact ProductSafety@springernature.com

Printed and bound by CPI Group (UK) Ltd, Croydon, CR0 4YY

30/09/2026

02225259-0007